2
Husqvarna
The Early Years

By
Alex Askaroff

Sewing Machine Pioneer Series

Husqvarna
The Early Years

By
Alex Askaroff

Sewing Machine Pioneer Series

Husqvarna
Established 1689
Husqvarna Vapenfabriks Aktiebolag
(Husqvarna Weapons Factory Limited)

Husqvarna
'Makers to the Crown'

On The Road Series

There are seven books in Alex Askaroff's **On The Road Series**. They cover his working life around Sussex encompassing a world of stories from the ages.

Book One: Patches of Heaven

Book Two: Skylark Country

Book Three: High Streets & Hedgerows

Book Four: Tales From The Coast

Book Five: Have I Got A Story For You

Book Six: Glory Days

Book Seven: Off The Beaten Track

"If you read any of Alex's 'On The Road Series' you will read them all. They are totally addictive, beautifully crafted and wonderfully inspiring."
Eliza Cooper

ALL RIGHTS RESERVED

The rights of Alex Askaroff as author of this work have been asserted by him in accordance with the Copyright, Designs and Patents Act 1993.

No part of this publication may be reproduced or transmitted in any form. No part of this publication may be stored on a retrieval system or be transmitted in any form or by any means without the Copyright owner's permission. The author asserts his moral and legal rights to be identified as the sole author of this work.

All images are by permission of the owner, in the Sewalot Collection, the Public Domain or out of copyright. If any image is incorrect please contact us for details to be included in further editions.

Amazon print for worldwide release 2022
Paperback ISBN: 9798813457340

Front cover image Alex Askaroff

© ALEX I. ASKAROFF

Dedication

I dedicate this work to all the enthusiasts and collectors around the world. Their effort, work and enthusiasm, however large or small, preserves our heritage, history and maybe a little of our souls.

Here I am with part of my Sewalot Collection in 2022. I have several Husqvarna machines in here which I dust and use regularly.

Foreword

Alex Askaroff grew up in the sewing industry and has spent a lifetime studying and writing about his craft. He is a world renowned expert on pioneering sewing machine inventors, creating Sewalot.com which has become the premier website for antique sewing machines with millions of visitors.

For decades he has assisted and consulted with novels, programmes and films, from the BBC The Repair Shop to The Great British Sewing Bee, The Singer Story, Made in Clydebank, to How The Victorians Built Britain with Michael Buerk and more. His expertise in the sewing field has helped countless enthusiasts and collectors, giving interviews and writing numerous articles for magazines and publications worldwide. Alex Askaroff has had Nine No1 New Releases on Amazon.

Please Note:
This complex academic piece can be skipped over at your leisure. Please use it as a rough guide to early Husqvarna machines and enjoy it in the spirit that it was written.

Index

Introduction........................ 9
Chapter One........................ 13
Chapter Two........................17
Chapter Three...................... 26
Chapter Four.......................30
Chapter Five.......................34
Chapter Six........................40
Early Husky Models...............42
Chapter Seven......................46
The Freja Sewing Machine........48
The Triumf Sewing Machine......54
Later Husky Models................60
Chapter Eight......................71
Chapter Nine.......................77
Nordic Sewing Machines..........90
Chapter Ten........................97
Chapter Eleven....................114
Chapter Twelve....................120
Chapter Thirteen..................123
Chapter Fourteen..................128

Introduction

Husqvarna have made some of the finest sewing machines on the planet. It was impossible to do a whole 'Sewing Machine Pioneer Series' without mentioning one of the best. It is also one of the most famous brand names in the world.

We cannot talk about the company and its brilliant sewing machines without first explaining how it came into being.

The very reason that Husqvarna exists is due to weapons. Although the Husqvarna name is now on a multitude of famous items, I will try and keep to their two main products that they successfully made for countless years, guns and sewing machines.

Finally, before we get started, a big thank you to everyone who sent in images from around the world. It has helped in bringing this academic research book to life. It is my hope that this first work will help and encourage others to take the flag forward, to correct and complete a more detailed volume on this extraordinary company.

The Husqvarna name in Europe and America became synonymous with quality and was on everything from motorcycles to outboard motors. I was once in the middle of a lake in Sweden when one of their ultra-reliable outboard motors seized solid due to weed. It was a long row back to shore. The next day the motor worked perfectly. No, that's not me in the boat!

I will occasionally mention other Husqvarna products just to try and keep their complex timeline in some sort of order. Please forgive any inaccuracies in the Husqvarna story. It is a complex and ever changing one.

You know something astounding! In the 13th Century you could have purchased an axe made from the finest Swedish steel from craftsmen in the Husqvarna area. Farmers would have bought cutting axes, stonemasons their pickaxes and Viking marauders their throwing axes. In 2022, as I finish my book, you can pop online and still order a brand new Husqvarna axe. How about that!

Husqvarna as a company and brand name is unique. They seem to have almost singlehandedly kick-started the Industrial Revolution in Sweden. They employed countless thousands of people and changed not only their area but the whole country as it prospered. They are one of the hidden treasures of Sweden. I'm amazed that no one has written about them in detail before.

Incidentally, if you ever want to see how I build these research books for the 'Sewing Machine Pioneer Series' I explain it all in my book The History Of Frister & Rossmann Sewing Machines. It is a fascinating insight on how to do it.

Hang on to your hats everyone, we are about to go on a rollercoaster ride through history.

CHAPTER ONE
The Beginning

The early years of Husqvarna are as fascinating as they are extraordinary. Certainly their history is unique in the sewing machine field. It is a story just waiting to be told. Please do excuse my spelling. Some Swedish words seem to translate several different ways. It has turned the last few remaining dark hairs of mine, grey! Also don't forget I spell in United Kingdom English which can be slightly different on a few words.

Now, for around the first two centuries, as Husqvarna grew, it was a major Swedish weapons manufacturer. To understand how and why Husqvarna began making sewing machines we are going on a journey through history and into the Swedish woods, way back before the official start of the company.

We have to go back to the beginning of their journey which officially started in 1689 but actually pre-dates that by a long period. I am going to explain how one of the world's greatest sewing machine makers began.

Firstly, I'm always getting corrected by people on how to say Husqvarna. I often feel like a school kid being told off. Husqvarna, both the town and the brand name are apparently pronounced Huus-k-varna. No letters please.

The Husqvarna Company officially started life as a Swedish arms manufacturing company in the town of Husqvarna by Lake Vättern, (the second largest lake in Sweden) roughly half way between Malmo and Stockholm in Southern Sweden. Today the town is spelt with a K rather than a Q. I will get to it later. It currently has a population of around 25,000.

The company name of Husqvarna comes partly from the position of the factory alongside the impressive Husqvarna River and the impressive falls where great salmon leap upstream in the spring floods. Husqvarna roughly translates to 'the mill house' or 'the mill by the house' and may have originally been, Kvarnhus.

The mill houses on the River Husqvarna would grind grain and others power the blacksmiths bellows, plus a thousand other jobs where unlimited power was needed. They also proved to be ideal for weapon making.

Water, draining from the low lying farmers fields flowed into Lake Vättern and from there down the rivers. It was this perfect position of land and water that led to the first mills being constructed.

Farmers used the power of the river to grind their grains for centuries and as more mill-houses grew around the area and the name Husqavrna was established forever (well until the town spelling changed, apparently around 1810, and the Q became a K).

Husqvarna to Huskvarna

Husqvarna Vapenfabriks Aktiebolag. Husqvarna Weapons Maker Limited, Huskvarna, Sweden.

This may be a good time to quickly explain the spelling of Husqvarna and Huskvarna as shown above. Simply, when the town changed its name to Huskvarna the Husqvarna Company (that had been trading across the world) decided to keep the original spelling, and has done to this day.

That simple bit of information took me a week to track down, crazy or what. With Google today I expect I could have done it in an hour or less!

Anyway, it would later be these mills that sparked the industry that provided the power to the factories before steam and electricity came along.

It was these very mills dotted around the countries landscape that helped power Sweden's Industrial Revolution.

CHAPTER TWO

The Husqvarna name (that appears on so many products) can rightly claim to have belonged to one of the oldest companies in the world.

Husqvarna originally was the arms factory to the Swedish Royal Family. It was in business for centuries before it started making sewing machines. Early Husqvarna history is sketchy at best and I have travelled the length of Sweden looking for details. Here is what I have discovered so far.

To begin with the company seemed to have evolved out of a need for military items in the medieval period around Jönköping just to the west of the town of Husqvarna. Today Jönköping is a thriving tourist and university town on the southern shore of Lake Vättern. It is the main city in Jönköping County which is the highest county in southern Sweden and perfect for a stronghold (as we shall find out). For as far as written records go back Jönköping has been a heavily wooded wild area. The main economy being animal and grain farming.

During the 1300s there was a well maintained fortress built by the Swedish King to replace a crumbling castle. It was called Rumlaborg. Its remains are still visible near Jönköping and Huskvarna today. Jönköping had gained its town privileges from King Magnus III in 1284 after it grew from a handful of small hamlets into a bustling town. King Magnus spent much of his time in his fortress on an island in Lake Vättern. Records show that the name Kvarnhus or Husqvarna was also regularly used to describe the area of mill-houses beside the waterfalls known then as Husquernen Falls.

All over the world the first regular use of free energy was from watermills. Simply build your water wheel and you have an unlimited supply of power 24 hours a day. As you can see here at Groudle mill, the wheel could turn anything from a row of sewing machines to a loom, or just grind flour.

Swedish villages starting with 'HUS' can often indicate a fortified encampment originally belonging to the crown. Even at this early time, the largest mills were run by the nearby Rumlaborg fortress to manufacture weapon parts, possibly axes and arrow heads, hammers and other tools, even armour.

Rumlaborg was an important stronghold for the Swedish Crown during the Late Middle Ages and, as it grew, attracted other weapon makers to the area. The 'Jön' from Jönköping means brooks, streams or creeks.

Interestingly, mills had so many uses besides grinding grain, they were the life blood of early industry, providing endless power and movement. Some mills, like the Valke Mill, were used to press the soldier's clothes! A few years later the same mill was converted to produce and grind gunpowder. As more tradesmen came to Husqvarna and Jönköping, Gustav Vasa gained permission for the Franciscan monastery to be rebuilt as a fortress to protect this crucial industry.

Under orders from the King the monastery was converted into a military stronghold for the area surrounding Jönköping. Its orders were to protect and secure trade and travel. Both Jönköping and Husqvarna were under constant threat from Danish plunderers on their raids.

Converting the monastery and its surrounding area into a military stronghold was a stroke of genius. It gave protection to the blacksmiths and tradesmen. Some of them were initially suspicious of putting all their skills into one small area. However with the added bonus of protection from the Crown they flocked into the area around Jönköping.

The military stronghold needed weapons and other supplies and now most of these could be forged by the local community of craftsmen.

Over the next few years the community seemed to slowly combine and join, growing into one distinct large industry from the multitude of small businesses.

Now we come to King Gustav II (1611-1632). He is still credited for the rise of Sweden from a rural farming country into a great European power. The Swedish King eventually became known as Gustavus Adolphus the Great. That can't be bad!

In 1620 he ordered all the rifle, musket and small arms factories to be organized into a number of Swedish communities. Jönköping was already growing (from its origins of a few forest huts, forges and small private businesses) into a thriving area.

Initially the majority of the workforce, centered at Jönköping, were living in assorted hamlets around Småland and Västergötland. That slowly changed as lots of small industries combined into one business, that of weapon making.

The Husqvarna Falls near Jönköping on the southern end of Lake Vättern. They were one of the most famous waterfalls in Sweden, often called 'the foaming falls' as they split into eight separate waterfalls. As the river thundered down it created a spectacular rainbow coloured mist that visitors loved. Cleverly, as the industry grew they protected the beauty of the waterfalls

Here you can see the Husqvarna Falls cascading by the later Husqvarna factory. It remained as a weapons workshop for decades, harnessing the power of water. This is currently the entrance to the Husqvarna Museum which opened its doors in April of 1993. I talked to Steven at the museum just before I went to print and found that they had recovered well since the worldwide Covid pandemic that started back in 2019. The museum is open regularly from 10am and is well worth a visit. Image Husqvarna Werksmuseum. Part of the river is now diverted for hydroelectric power.

Here is an early view of the Husqvarna Works long before the museum opened there.

Weapon manufacturing requires huge amounts of power, which in our case was supplied by water or hydropower. Unfortunately due to silting up the Jönköping water, their normal supply had become unreliable and a better source was needed.

Now here our history divides. Some tell me that when the river, and consequently the mills at Jönköping failed, it was orders from the King of Sweden, Gustavus Adolphus, that was responsible for the moving of arms manufacture to the Husqvarna River.

Some say it was actually Count Erik Dahlbergh many years later in 1688. He promoted the idea to King Charles XI who approved it. Only time and the fast improving Internet will tell.

Count Erik Jönsson Dahlbergh (1625-1703) was a fascinating character. Orphaned at an early age, after countless jobs and adventures, he became a military engineer. He served on campaign in the Polish Wars, rising to 'engineering adviser' to the King.

Although an excellent architect and engineer, his speciality was military fortifications. By 1676 he had become director general of fortifications for the Swedish Crown. He became friends with not only King Charles X but his son Charles XI. He rose through the ranks becoming Field Marshal Dahlbergh. Through his diligence, hard work and service he worked his way up to become part of the Swedish Nobility.

It was probably on Erik's advice that the King ordered the construction of improved mill works and forges at the waterfalls just outside Husqvarna. The work was completed in 1688. The area now met the requirements for reliable power and increased weapons manufacture. Erik Dahlbergh was also the manager at the state arsenal in Jönköping at the time.

At Husqvarna, they had ample supply of hydropower and charcoal from the forests to heat their forges. So the armoury was slowly moved from Jönköping to the new power mills at Husqvarna and our historic tale begins.

Interestingly, just like here where I live in East Sussex, England, the gunpowder rivers and tributaries have all but dried up around Blackboys and Heathfield, due to lack of maintenance. Exactly the same has happened to some parts of modern Huskvarna and just old river beds can be seen where once water flooded through to giant mills.

Can you imagine for a second the clearing of the huge ancient forests? The laying down and building of great mills, the hundreds of men needed to work, clear, cut and chop. People cooking and feeding, looking after the workers while they lifted great beams and wheels into place.

And then finally the starting of the mills that would bring unlimited power to their new armoury. Endless fuel to drive the forges, power hammers and billows. All this actually happened and a new chapter in Swedish history was born. The Swedish Crown, the government, had created an industry, a community and a town that would prosper for countless generations.

CHAPTER THREE
Jönköping Rifle Company

By the 1680s war was smouldering with Denmark and a successful arms manufacturer, Bengt Larsson Billing, was increasing production at Jönköping. In 1688 relations with Denmark were at breaking point and war seemed inevitable with Sweden. Even better for Bengt, the troops that were to march off to war to fight were billeted around Jönköping. Even the mobile artillery were stationed there.

Weapons training happened daily with the soldiers. They were using what we now call slow fire flint dragoon muskets. They were also trained with flint hand muskets and wheel locking carbines.

The pressure was on to finish the King's latest armoury to supply badly needed arms for the war. In 1689 King Charles XI was informed by special courier that his new factory and Royal Armoury was completed at Husqvarna. It was officially opened the same year. Some say it was the Governor of Jönköping and founder of the company, Count Erik Dahlbergh, who personally made the long trip on horseback to inform his king that increased musket making had begun (the King was at Stockholm Castle at the time).

From 1689 arms production continued for the next three hundred years. There is some great info later in our tale about the last 15 shotguns ever produced at Husqvarna.

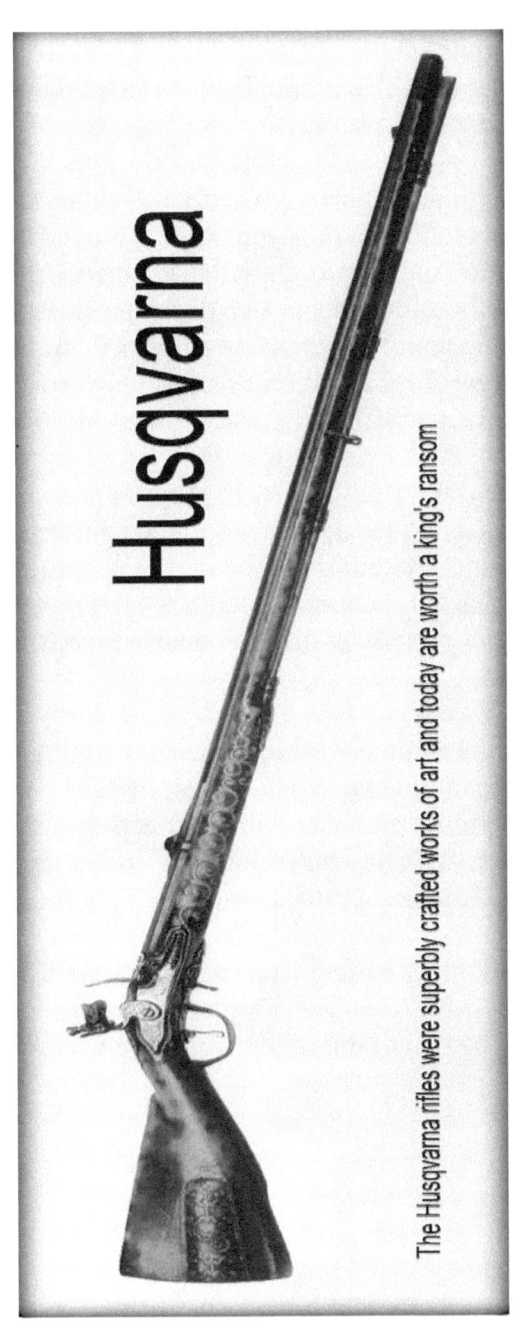

Husqvarna

The Husqvarna rifles were superbly crafted works of art and today are worth a king's ransom

Although the roots of Husqvarna pre-date 1689 it is 1689 that is known by the Husqvarna brand as the date that they were officially founded.

The combining of services and industry led to a boom in manufacturing (long before the Industrial Revolution in Britain). Initially the small armoury was hand-building about two dozen muskets a week. Amazingly output rose at the new mills in a single year from 1,100 rifle barrels to over 12,000. The Jönköping Rifle Company was booming.

Because of this new ability to produce arms in large numbers, later that same year, the King ordered his new armoury to build muskets for the entire Swedish army. I believe that the royal 'H' would have been proudly marked on their weapons from this point.

Big orders and the wealth that came with it, allowed the company to expand and invest in its area, building more premises and constructing dams at Husqvarna (Huskvarna) to provide continuous water to feed the arms factories and foundries.

The Jönköping Rifle Factory ran successfully for the Swedish Crown for many years until eventually being renamed to the Husqvarna Rifle Factory.

Husqvarna were proud to be associated with the Swedish
Crown and used this logo to enforce their royal patronage.
This is from a sewing machine lid.

CHAPTER FOUR

Husqvarna Gevärsfaktori (Husqvarna Rifle Factory)

In 1757 the Swedish Crown sold its stake in the business. For the first time it became a private company trading out of its assembly plants around Jönköping and Husqvarna.

At the same time the company name was changed into Husqvarna Gevärsfaktori (Husqvarna Rifle Factory).

By 1850, almost all activity in rifle and musket manufacture around the area had been moved to Husqvarna and its purpose built foundry factory which employed over 1,000 people.

In the same year the arsenal was privatised under the ownership of Fredrik Ehrenpreus. The Ehrenpreus family then ran Husqvarna.

They installed new plant including the latest drop hammers and specialist forges. Improvements in manufacturing mechanisation and machinery brought in by the Ehrenpreus Family meant parts could be interchangeable instead of each piece being hand finished. For the first time in Swedish History true mass production was taking place.

Interestingly it was much the same in America where the need for arms pushed the boundaries of mass production. In the 1850s, Isaac Merritt Singer had many famous visitors to his engineering works including Oliver Winchester and Samuel Colt. They all wanted to see how Isaac could mass produce sewing machines to such exact tolerances.

In 1867 Husqvarna became a Limited Company. On the 22nd of November 1867 the first AGM was held at the Husqvarna Small Arms Company.

The first serious mention of Husqvarna as an individual company starts in 1869 with a big military order for weapons (but that was short lived as we shall find out).

For nearly 200 years the company produced mainly armaments but things were about to make a dramatic change.

Around the early 1870s there was an economic depression in Sweden and drop in military orders. There were few major wars in Europe. Peace meant low orders for weapons. Husqvarna diversified into shotguns and sporting rifles to supply the demand for accurate hunting rifles in Sweden. However, to survive they needed to come up with something fast as things were going to take another sudden and unexpected downturn.

Interestingly, hunting is still a hugely popular pastime in Sweden with around 300,000 people regularly hunting in the vast wilderness. Everything from reindeer to wild boar are on the menu!

Things came to a head for Husqvarna when the Swedish Crown cancelled its long running standard orders for armaments. To survive Husqvarna needed new ideas and fast. By now countless people depended on the company. The whole community was affected and it looked like an economic disaster was looming.

You have to understand that the surrounding area, mainly due to Husqvarna, had moved away from being a mainly rural farming community. Over the centuries, as the thatched huts turned into hamlets, then small communities and villages, Husqvarna and Jönköping grew in prosperity.

Men whose fathers and grandfathers were farmers, living in secluded settlements, were now young skilled craftsmen with a trade. Husqvarna had grown into a bustling town, all centred around skilled employment, mainly from the Husqvarna factory.

If that industry failed in the 1870s, Huskvarna and its surrounding area would be devastated for generations. This sounds implausible but that is exactly what has happened around the world in countless places.

One of the worst in my lifetime was the closure of the great Singer factory in Scotland. It coincided with the loss of shipbuilding in Clydebank and it took 30 years to recover. Films from the 1970s in Kilbowie and Clydebank made the area look more like a war zone.

During the 19th Century in America, Britain, Germany and elsewhere, new factories were popping up like mushrooms on a warm autumn evening. The Industrial Revolution had spread worldwide and was in full swing.

On closer inspection many of these factories were making a new gadget that mechanically and almost magically joined material together.

More importantly it looked like almost every household on the planet wanted one. They all wanted a sewing machine. Enter into our story, Wilhelm Tham.

CHAPTER FIVE
From Weapons To Plough Shears
Wilhelm Tham

The Husqvarna Museum in Sweden supplied some great information here on how Husqvarna managed to survive its largest struggle. Let's look at who we have to thank, Wilhelm Tham.

With Wilhelm Tham at the helm of Husqvarna the company looked to the booming economies of the world for the latest ideas.

It did not take Wilhelm Tham at Husqvarna long to come up with plans to produce a sewing machine.

Switching some of its production to sewing machines took two years, with the foundry finally coming online in 1872.

Wilhelm Tham was promoted to president of Husqvarna in 1877 by the entrepreneur and industrialist Hugo Tamm. Hugo, who had a large stake in Husqvara, specialised in taking ailing companies and resurrecting them.

At the time the company had 170 workers. Wilhelm understood marketing. For example when the Husqvarna Freja was produced, each early machine came with a certificate signed by Wilhelm himself personally assuring its quality.

This very unusual certificate circa 1900 was kindly sent in
by Victor Joahnsson.

Wilhelm Tham understood exactly that by diversifying Husqvarna would become stronger. The more items it diversified into, the bigger the company could become and the better it would be at withstanding economic downturns. By diversifying the company would not need war to prosper!

Husqvarna would start with sewing machines but go on to produce everything from bicycles to motor cycles, kitchen ranges, and cast iron pots and pans.

Later they expanded into power saws, chain saws, electric stoves, fridges, freezers, washing machines, lawn mowers and much more. It seemed that the company had the ability to put the Husqvarna name on just about anything that you could think of and sell it.

Their Husqvarna wood burning stoves proved very popular for cooking and heating. They had huge efficiency savings over a traditional ranges and open fires. With the cost of fuel today they will probably come back like a rash!

Husqvarna carried on making arms, sporting guns, pistols, rifles and their premier shotguns. However it would be their sewing machines that really prospered over the next 100 years.

Interestingly it was their expertise in handling the small complex parts in guns and rifles that made the manufacture of sewing machines the perfect new product.

Wilhelm Tham guided the company through its most tempestuous times. It could be claimed that it was due to this one man that the company not only survived but eventually became the world renowned name that it is today. He invested heavily in the town its church and school. He seemed to be the catalyst that built the roots that modern Huskvarna stands on today.

Before long the company was on a boom, making just about everything for the household.

Diversification away from just producing arms brought many new products to the company and a few surprises. The Husqvarna Reliance ice-cream maker was a big hit across Europe but had a limited seasonal market.

Let me just finish the chapter on the Wilhelm Tham by telling you that when his son Gustaf Tham retired in 1946, Husqvarna had gone from 170 workers to over 6,000. Wilhelm and Gustaf had guided Husqvarna through their most tumultuous years and turned them into a global phenomenon.

It wasn't just Husqvarna who diversified to survive. Businesses all around the world switched from firearms to profitable peacetime products. One of the most famous here in England was BSA Motorcycles. BSA or Birmingham Small Arms have made some of the most famous and iconic British motorcycles. A drop in small arms demand in the 1880s saw BSA switch, first to bicycles, and later to motorcycles. It was only during the First World War that BSA switched back to arms manufacture for a period.

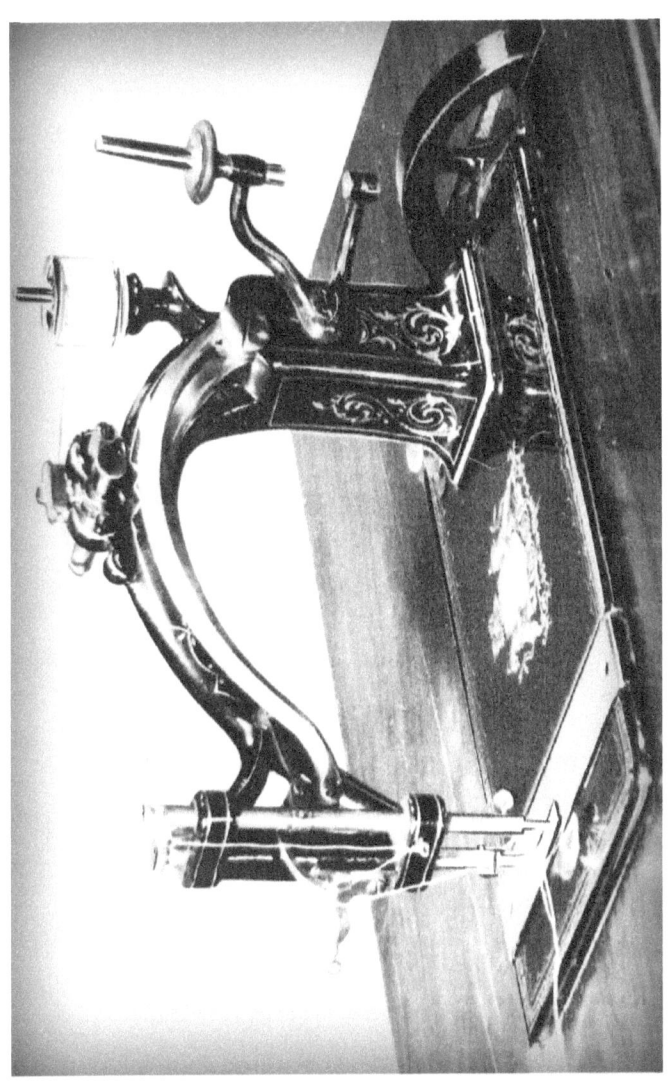

1872, the Nordsjernan, North Star or Northern Star was
the first and most beautiful sewing machine that
Husqvarna ever produced. Husqvarna Museum, Sweden.

CHAPTER SIX
Sewing Machines
1872

At last we are here. I feel like doing a happy dance. Years of research can be tucked away and now we can get on to my true passion, sewing machines. If you skipped the first chunk of this book and went straight to sewing machines then go and stand in the corner and try to look guilty☺

As the 19th Century progressed sewing machines, that had started life in the 1840s, as a useless gimmick, had become the latest 'must have' craze.

Machines were selling like hot cakes around the world and Husqvarna, a company loaded with skilled staff who were used to forging steel, working hot metal and precision engineering, took to them like a duck to water.

Once their selling techniques caught up with the competitions, sales boomed. But not before a little hiccup! The Northern or North Star. Let's call it a learning curve.

Their first sewing machine was one of the most beautiful sewing machines ever made, the Northern Star or North Star. A simply stunning machine. Unfortunately, it did not stitch well. It was too delicate, had flaws and was soon updated.

The Nordsjernan or North Star. This model became fondly referred to as the 'cat back'.

The Husqvarna North Star sewing machine is one of the most sought after sewing machines. Few people have seen a real one. I have even scoured Sweden looking in every antique and sewing shop I came across with no luck.

The prototype Husqvarna North Star may have been a failure in sewing terms (as the arms giant learnt a new trade) but what a beautiful looking machine!

Only about 500 of these beauties were made. To make things worse for collectors, Husqvarna recalled many when they realised it would hurt their reputation for excellence, offering a part exchange on their new improved model.

There may only be a handful that survive to this day. The Husqvarna Museum in Huskvarna have a beautiful treadle on display. I wonder if I tried to slip the head into my rucksack I would promptly be escorted from the property!

Early Husqvarna machines are rare. From 1872 until post WW2, Husqvarna's main market was Scandinavia. Around the world and even in the United Kingdom, early Husqvarna machines just don't turn up. To make things worse several models were sold with no makers badge. A few were just marked, 'Made in Sweden'.

All Husqvarna's early machines seemed to have had names until Husqvarna started to add model numbers. This is what I have come up with to date. There are probably a few unknown Husqvarna models still out there?

Early 'named' Husqvarna models

The Nordsjernan, 1872-1874
Family Favourite, Howe-Weed system 1874-1877
Family Fiddlebase 1877-1883
3/4 size shuttle (Singer licence) 1878-1883
Svea, female warrior, Valkyrie 1877-1892
Singer medium copy 1878-1886
Freja, in several designs 1877-1925
Salon A6 fiddlebase on iron bed 1878-1883
Paw Foot G&B 1889-1904
The Gota 1890-1892
Idun (goddess of youth) 1892-1896
Ingeborg (fertility goddess) 1896-1902
Husqvarna VS 1890s-1959
The Dextra (skilled) 1898-?
The Triumf vibrating shuttle 1885-1931
The Nordic, Comrie & Scandinavian 1912-1970

The next machine Husqvarna produced was a much improved model using many ideas where the patents had expired and some designs under patent licence. By combining the best of what was around in the 1870s they produced the Family Favourite.

I believe that they may have bought the rights to produce the American Weed, Howe and Grover & Baker designs under licence in an effort to come up with a super reliable model. In 1854 T. E. Weed had patented a mechanism that was consistent and produced a good lockstitch. Howe and Grover & Baker ideas were also used. I have written extensively about them in their own books in my Sewing Machine Pioneer Series.

The Weed mechanism was sold out under licence to several European makers. The next image is an almost direct copy of the popular Weed sewing machine of the day and sold well in Europe, especially Germany, until it became outdated and phased out at the beginning of the 20th Century.

In America the same machine was more commonly seen in a treadle but in Europe the 'paw foot' hand cranks models in various designs were popular.

The Husqvarna Grover & baker design is super rare today and was in production from 1889 to 1904. There is a spectacular one in the Husqvarna Museum, Hakarpsvägen, Huskvarna, Sweden.

The Paw Foot Husqvarna Model A (Howe/Weed/Grover & Baker design). This design was sturdy and reliable. It also had a Germanic appearance that the Swedish found familiar. In fact several German companies including Haid & Neu and Wertheim made almost identical sewing machines for a few years. Husqvarna produced this model from 1889-1904. Below is the super-rare Husqvarna Singer Salon A6 sent in by Victor Joahnsson. Notice the cast H.

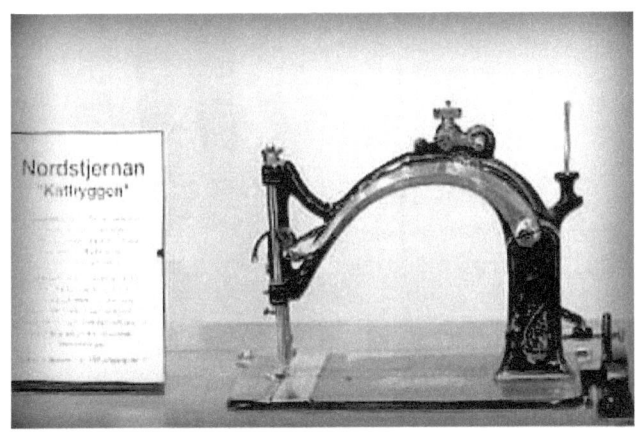

The two rarest Husqvarna sewing machines are the North or Northern Star and the Husqvarna Grover & Baker copy (below). The North Star snapped on a visit is the actual one in the Husqvarna Museum, kindly sent in to me by Scott Geller. They are so rare that I have not come across either machine for sale in over 50 years.

CHAPTER SEVEN
Husqvarna in Britain

The Husqvarna machines were first imported into Britain by the American Sewing Machine Company (who were in fact were not American at all but simply imported mainly American machines). The A.S.M.C. American Sewing Machine Co was founded in 1863. Because all the best sewing machines were coming from America it was a clever marketing strategy.

E. Todd, sold both styles of 'The Champion Of England' and imported the Husqvarna Freja sewing machine model as early as 1886. He sold them to stores like Reed's in London. Todd and the A.S.M.C. were somehow tied up together but I have yet to discover how. Probably agents and importers.

This needle plate from an 1880 Husky Freja sewing machine holds a host of information. Legend goes that the seven stars marked the seven states of the Confederacy that held out against the Union in 1861. It could be just a legend but feelings ran high after The Civil War in America and products that were for the Southern States from a neutral country sold well.

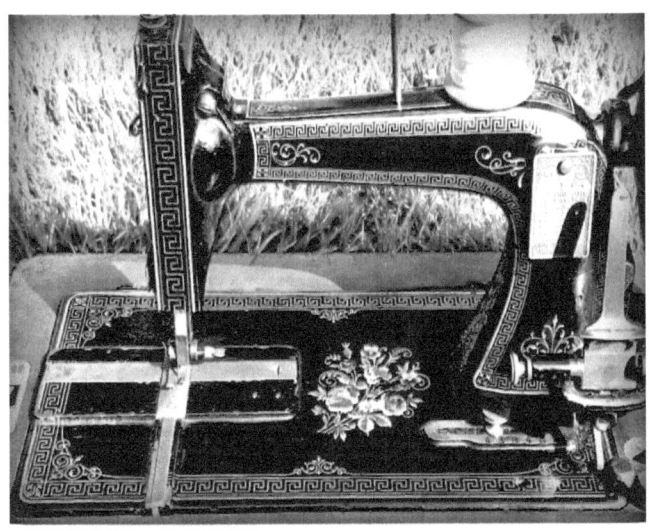

Here is a machine that E. Todd sold via the A.S.M.C. The American Sewing Machine Company sold some beautiful machines in Britain. This could be a Swedish Husqvarna or a German machine. The A.S.M.C. were very careful to remove all markings and badges before selling machines. This may have been due to import duties. High taxation from certain countries made importing their products unviable.

The 1870s & 80s were the pinnacle of beautiful machines. Endless hours were taken in decorating sewing machines to make them look stunning. Many even had Mother of Pearl decoration and the shone like jewels. All this work came at a cost and by the 1890s, with competition coming online across Europe, these labour intensive decorations were dropped. This machine is in my Sewalot Collection. I must get her out and run up a few stitches.

The Freja Sewing Machine
1877-1925

This is the Husky Freja 'High Arm' in my Sewalot Collection (and on the front cover). This design is based on the bestselling Singer 12. It was extremely reliable and popular. It sold virtually unchanged for over 30 years.

By the 1880s Husqvarna produced their first really big selling machine, the Husqvarna Freja. This was pretty much identical to many of the German imports, which in turn based their machines on the bestselling Singer transverse shuttle of 1865. The first Husqvarna machines all had names. It would be many years later that Husqvarna started using the Viking name as standard across their machines.

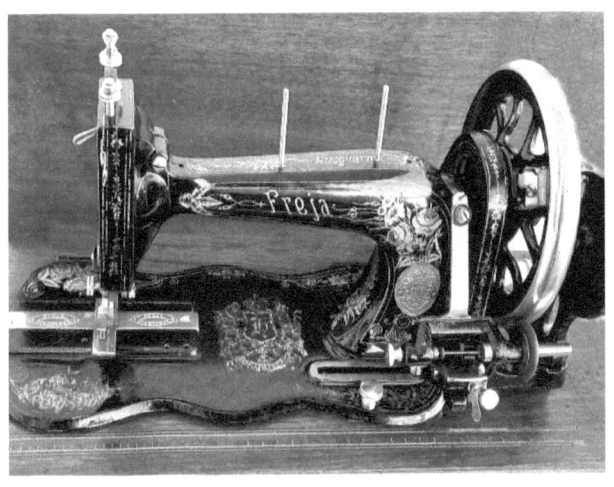

The first Freja A6 was the early ¾ size transverse shuttle known as a fiddlebase due to its curving bed, circa 1880. During this period there were constant changes to the shuttles, hand wheels and other parts.

As orders flourished so did the factory and the surrounding area of Jönköping and Huskvarna. More and more young workers left their traditional jobs as field hands and farmers to become skilled engineers and technicians.

The first Freja were fiddlebase models. So called because of the curving shape of the bed. A copy of the bestselling machine of the period, the Singer model 12. Very few of these survive today. This beauty is owned by Hans C. Endrerud, circa 1880. It uses a MY1014 needle.

Early Husqvarna sewing machines are rare. The most likely model to turn up is the Husqvarna Freja which first started production around 1877 and changed shape over the next 30 years.

The magnificent Freja model was based on the mighty Singer 12k transverse shuttle model of 1865 but with a more Scandinavian high-arm square look.

The Freja uses a 12x1, 13x1, 331 or MY1014 needle. Remember that the long groove faces towards the operator and you thread front to back. Much like most modern sewing machines today.

The later Freja was another transverse shuttle but on a straight bed, again it was not like the larger 'high arm' machines but ¾ size.

For decades the Husqvarna Freja was the company's bestselling sewing machine and earned Husqvarna a reputation for fine sewing machine engineering as well as beauty.

It was these machines that helped to establish Husqvarna, not only as a weapons maker but also a maker of high quality domestic goods. It was the first step that would lead to many diversifications over the coming years. There was only one rule at the Husqvarna plant, quality, quality, quality.

The company would go on to patent many of their own superb ideas. Possibly their best idea was the perfecting of steel that was self-oiling. This alone revolutionised machinery worldwide. We will get to that a little later in our story.

Husqvarna Freja 'high arm' had several different decals over its production lifetime but was still a transverse shuttle machine first developed in the early 1860s.

Often, when the Freja model was sold, all that was on the top was **Made in Sweden**. Values on these rare machines are hard to say. There is a growing collectors market, especially in Sweden. They still sew better than many new machines and look utterly beautiful.

Freja, can be spelt many ways such as Freyja or Freya. She is the Norse Goddess of fertility. Even today she is one of the most popular Scandinavian names for girls. We know her each week for Friday is from the old Viking word for Freja's Day.

Husqvarna Triumf

Husqvarna also made the Husqvarna Triumf sewing machine. By the 1920s most manufacturers machines looked similar but the Triumf was special, a rollover from the last century and very popular.

Note the interesting tension calibration of the Husqvarna Triumf sewing machine. The 'high arm' Triumf was made from 1885 right up until 1931.

Husqvarna classed it as a 'professional model' though in truth it was a simple shuttle machine with a larger sewing area. The Triumf may have been one of their first shuttle style of machine but moving straight forwards and back, unlike the later VS machines where the shuttle swept in an arc and missed less stitches.

The Husqvarna Triumf was a stunning machine. Once again the models varied during the long production run. This superb machine is owned by Hans C. Endrerud. Production of the Triumf was from 1885 to 1931. This was one of the last opulent Husqvarna models that hailed back to the 19th Century. Its Art Nouveau classic lines are a sight to behold.

The rare Husqvarna Dextra hand crank sewing machine was basically the same as the Freja but with different decals. They are unusual and hard to find in nice condition today. The Husqvarna Dextra was also a high-arm German-design transverse shuttle of the late 19th Century.

This is the Husqvarna shuttle, a copy of the Singer 17 with a few minor alterations. Some say, due to exports, this model ran (in various different colours and decals) from the late Victorian period right up to the 1950s.

This Husky VS or vibrating shuttle is far rarer than it looks and I have only seen a handful in over 40 years. This beauty was sold by the Rothenborg Agency in the capital of Denmark, Copenhagen. It was in my Sewalot Collection until I sold it in 2016.

The VS model above sold for almost three decades with little changes. The later WW2-era models had far more austere decoration unlike the beautiful and opulent 1920s and 30s decals as seen here.

This is the super-rare Husqvarna Svea owned Gisela Mathias-Fröhlich. Interestingly it was Gisela who bought my beautiful VS on the previous page. Svea was the female warrior from Norse legend. As a Valkyrie and Handmaiden she would guide the souls of the deceased warriors to Valhalla. She is one of the most patriotic Swedish symbols.

By 1902 a new era of sewing machines was dawning. The main 'central bobbin' patents that Singer and other American patent holders held were all running out.

Husqvarna and others were free to make round bobbin or CB style machines. The factory tooled up and from around 1903 they started manufacture of their first round bobbin model. The CB Husqvarna machines ran for decades with few major changes.

The Husqvarna CB was the first of the modern era machines. Amazingly the central bobbin Singer 15 style of machine is still widely made today around the world. It was tough, basic and did a good job.

The Gota and the Ingeborg sewing machines may have used the Wheeler & Wilson bobbin system. However they are so rare I have yet to see one close up. When I asked several online vintage enthusiasts sites around the world, no one came up with either. How rare is that!

Some Later Husqvarna Models
CB (central bobbin) 1903-1934
CB X & XI (central bobbin) 1904-1951
CBN-12 1934-1958. Sold in Egypt as the Nefertiti
C III 1908-1912
KM 1912-1917
CL 10 1913-1959 VS. Its extended life due to exports to Africa and South America
CL 18 1955-1957
CL19 ZZ 1959-1966
CL 20 1953-1955
CL 21 1955-1966
CL 24 19..-1966
CL 25 1924-1947
CL 26 1926-1947
CL 262-263 INDUSTRIAL 1952-1966
CL 27 1935-1966
CL 28 1935-1942
CL 32 & HR 1939-1955
CL 14 ZZ 1939 OR 44-1954, aluminium.
CL 33 1947-65
CL 44 1946-1966
CL 45 1950-1964
CL 49 (1301 ZZ) 1961-1967
CL 51 1957-1967
CL 7 1958-1965
CL 8 (0201) 1959-1967
CL 38 (8311-10) 1964-1967
CL 71 1958-1966
CL 2000 1961-1966

The modern era of sewing machines had arrived for the company with their near perfect Husqvarna (Singer model 15) sewing machine which they named the Husqvarna CB X & CB XI. The first central bobbin machines ran from 1904 right up to 1951 with minimal changes.

The Husqvarna Museum in Sweden has an amazing story about a Husqvarna CB XI that is well worth repeating.

In November of 1918 the steam ship Per Brahe was on its way to Stockholm loaded with goods. The Husqvarna sewing machines and stoves were loaded on board with the help of a young Thure Andersson. The ship sank in Lake Vättern, everything and everyone was lost.

However, that is not the end of the story. In 1976 a few of the model CB XI sewing machines were salvaged from the bottom of the lake and returned to Husqvarna. Thure, who by now was a 73yr old retired sewing machine engineer, eagerly attacked the machines. He painstakingly managed to bring a few back to life. Astoundingly even though they had been submerged for over half a century they made a perfect stitch. Now that's a story.

One of the machines is still on display at the museum.

The first Husqvarna CB 'central bobbin' sewing machine.

Husqvarna moved with the times and by the 1920s they were producing a modern machine with an improved front tension and larger reverse.

Husqvarna continued to expand their range of products adding new items as the markets opened up. In 1895 they produced Sweden's first typewriter. By 1919 the company were manufacturing their own engines. They just kept going from strength to strength, always keeping their quality at the highest level.

Fancy a little laugh!

I once came across a Husky 2000 that had been dropped from a school table and thoroughly abused in its time as a school machine. Across the machine face was scratched, **I hate Dave**! Although dented and disfigured like an old boxer, the machine still made a superb stitch.

Husqvarna bought out its first electric sewing machine in 1934 and in 1935 they had their best sales year. Images on their sales brochure would now enter the modern world. There would be no more traditional almost Victorian sewing parlours like this one.

Husqvarna Central Bobbin CB machine. The Husqvarna Class 12 CB model was similar to the CBXI which first came out in 1904. This upper machine was sold in Holland and was a copy of the hugely popular Singer 15 which sold worldwide (the patents had disappeared long ago allowing anyone to use the mechanism). Mesje, the small word just visible on the front of the older one above translates to knife. A coat of black paint was added for the austere 1930s and 40s and now it was the CB-N which ran from 1934 to 1958. Many thanks to Patricia More Frey for the image.

The Husky 10 (not the Model X) sewing machine was still a vibrating shuttle 'old style' sewing machine but the machine now had a reverse. Later, from the 1930s the model 10 or CL10 came in a crinkle finish coat of paint. The Husqvarna vibrating shuttle models were the longest running machines. Due to excellent exports they ran from the Victorian period up to the early 1950s.

The final Husqvarna 11 & 12 was also supplied as a free arm or flatbed (as above) now marked Husqvarna CB-N central bobbin. Rather than update, Husqvarna added another colour coat of paint. The early 1950s was the end of production for the model. Many thanks to Susie Cordia for the picture of her Husqvarna 12 CB-N from 1954.

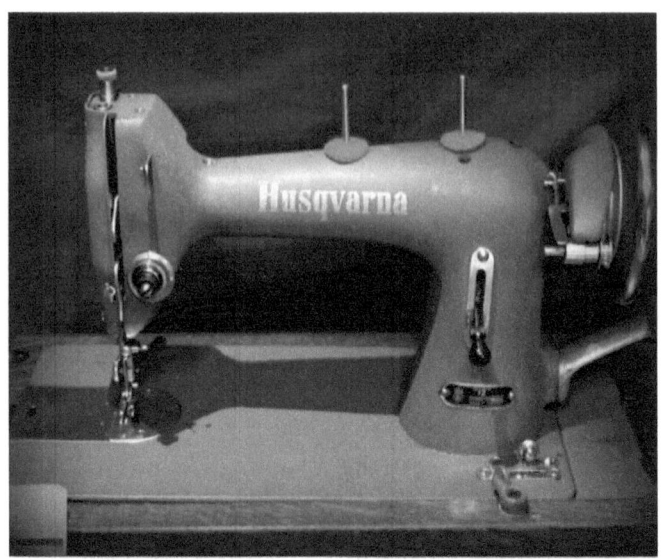

Here is another Husqvarna Class 12 CB-N. The CB-N was a near perfect stitch to compete with the Singer model 201. It could be sold as hand, treadle or electric. This model, kindly sent in Keith Hallgren, still make a beautiful stitch. The second cotton peg could be used to wind a bobbin as you were sewing or to add strength to your stitch by threading two reels of thread through the machine and sewing. You need a larger needle to do this to allow both threads to pass through.

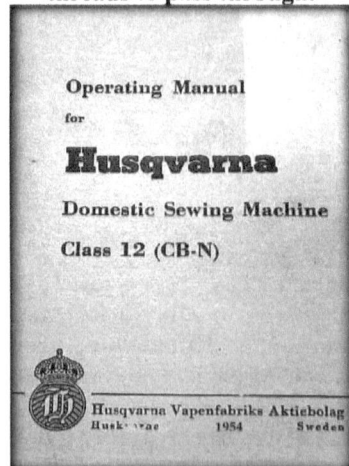

In 1934 they introduced an all-black Husqvarna CB-N a central bobbin model. It was a basic straight stitch updated 12 (based on the Singer 15). It ran for almost 25 years without hardly a change. By the late 1950s it was outdated and heavy. Similar machines are still made to this day around the world.

The Husqvarna HR was very similar with a different stitch regulator and ran from 1939 until 1955. It was slightly faster than the CB with a smaller lighter hand wheel.

By the end of WW2 there was a massive surge in production of sewing machines and the Swedish factory was soon producing over a thousand machines a week.

They must have had a brilliant research and development section at Husqvarna because over the next few years new ideas poured out of the business. It was like they had fallen over in a patch of stinging nettles. New machines, new ideas, new products. Husqvarna seemed well prepared for the modern world and the boom times that were ahead.

Incidentally in the 1930s Husqvarna was still labelling its sewing machines with Husqvarna Vapenfabriks or Husqvarna Arms Factory. During WW2 Husqvarna carried on with arms manufacture and I believe the last AK4 rifle ceased production in 1980. This was a common practice amongst sewing machine manufacturers. Their expertise in small precision parts made them ideal for arms and many helped with the war effort on all sides.

Amazingly Husqvarna worked to 1000th of a millimetre. That's a micro-metre or what's called a micron. If 70 microns were stuck together they would be the thickness of a human hair! I can't imagine working to those tiny tolerances.

I always remember one of my old masters who taught me precision sewing machine repairs saying, "Don't forget Alex, repairing modern sewing machines isn't rocket science. It's far harder than that!" I thought he was kidding. Silly me.

Many sewing machine factories, like most other factories, all helped during times of war. In my book on the extraordinary Singer 201, A Perfect Stitch, I go into detail of the arms and products that the great Singer factories made during WW2. Some of their handguns are now the most expensive in the world.

Interestingly, just at the outbreak of World War Two, almost at the same time as SAAB was starting to think about developing a small car, Husqvarna invested heavily in producing a 'peoples' car similar to the Fiat 500, Volkswagen Beetle and Citroen 2CV.

It was equipped with a German DKW engine but development was cut short due to mounting costs. I don't think it ever went into production.

By the 1950s there were over 100,000 sewing machines a year pouring out of the factory. In the mid-1950s Husqvarna shares were floated on the stock market for the first time at the Stockholm Stock Exchange.

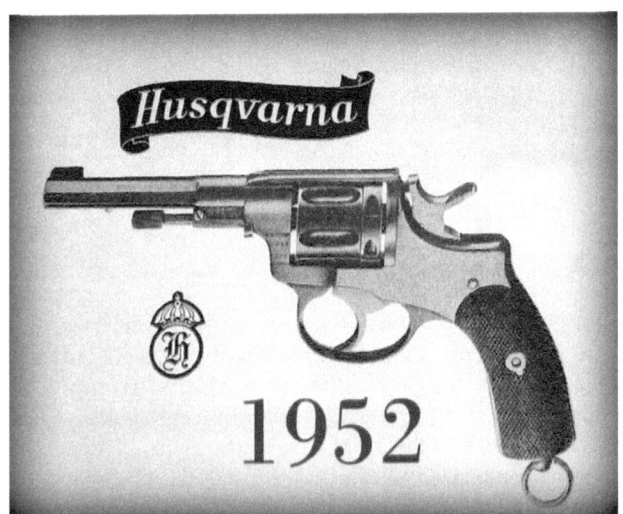

Diversification over the years was the key to Husqvarna's long life. No company had ever managed to constantly keep up and change with the times like Husqvarna. Studying the company would be a great university degree.

It is difficult to imagine the size of the Husqvarna works at its height from the late 1950s onward. This is only a fraction of the huge plant which by the 1950s employed thousands of staff. It's crazy to think that this area would have started as a few huts next to the river. Generation after generation of families around Jönköping and Huskvarna owed their livelihoods and prosperity to the Husqvarna Company.

CHAPTER EIGHT

Husqvarna to Viking
The Green Husky Years

From around WW2 up until the 1960s, Husqvarna developed a unique covering for their sewing machines. The closest colour today would be seafoam green. No other maker added such a final touch of luxury. These became known as the 'green years'.

Legend tells that when they were deciding on this fabulous colour it was after heavy rain. Water was cascading down the Husqvarna Falls by the factory works. The thundering mist was being caught by the sunlight streaming through the trees creating a beautiful sparkling green. It was this very metallic green that the company decided to try and replicate on their machines.

Interestingly there is a similar legend about First Nation American Indians and the Shenandoah River. One of the many legends is that the sparkling river inspired the name Shenandoah or 'daughter of the stars'. It may be true. Who really knows?

I love these little snippets. To this day no one is quite sure how Husqvarna perfected their super hard translucent metallic green coating.

Husqvarna was expanding worldwide but many countries were not aware of the powerful brand name. A decision was made to use the traditional Scandinavian name, Viking. Almost everyone around the world knew where the Vikings came from. It was a brilliant marketing move, so simple but totally effective.

The name was used more and more, some models used more emphasis on the word Viking and some less, depending on which market they were selling in.

As far as I am aware Husqvarna have not dug up any serial number records for their early sewing machines. Hopefully one day they will discover some.

A few years back I was contacted out of the blue by a lady who lived opposite the Bonnieres Singer factory. She had pulled out of a skip hundreds of prints that the factory had dumped prior to closing. She kindly sent me a whole bunch of irreplaceable leaflets and Singer designs. So who knows what may turn up? Anything is possible.

Dating a Husqvarna has to be done by machine, model and style more than anything. In fact I cannot find many of the names of their designers either, which is a shame as there are some incredible innovative features on many Husky machines. Hopefully they will come to light as our modern world progresses.

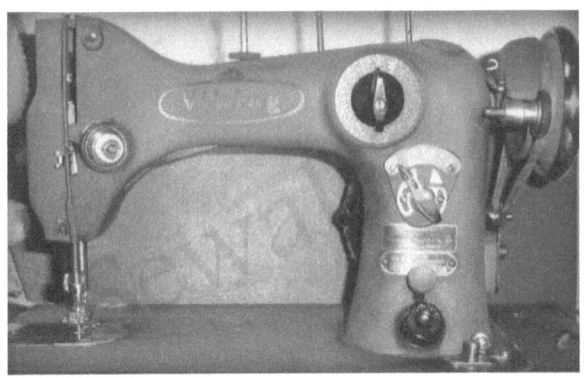

This is the Viking Husqvarna 33-10 a super smooth zigzag with a full industrial rotary hook. It may have been launched in 1933 but certainly the early 1930s. The Husqvarna name was also on several industrial lockstitch models set in their own large flatbed tables. The Husqvarna CL 27-20 was very similar to the Singer 196k. A perfect medium size industrial. The Husqvarna 24-10 was similar the smaller Singer 95k industrial. You can Google any of these model numbers to see the actual machines. The model 33 was powerful enough to be sold to factories with complete reliability. Amazingly there was even an embroidery unit in the back but I have never figured out how it worked.

Some say that the first alloy zigzag Husqvarna 14/34 came out before WW2, others say during, and some say after! It looks very much like the Elna machine of the same period. This model is currently on display at the Stockholm Technical Museum in Sweden. It has a built in light and that bar is the drop down knee power control.

By the middle of the 1950s the Husqvarna Company, from its early beginnings in the woods outside Husqvarna and Jönköping, had become a world renowned manufacturer.

Interestingly besides being the No1 sewing machine in Sweden, Finland was also a keen follower of the Husky machines. Only Singer outsold Husqvarna in Finland. This may have been due to the fact that until about 30 years ago around 10% of native-born Finns spoke Swedish as their Mother Tongue.

The Husqvarna Viking Model 18 sewing machine. The 18 was a brute of a machine and as heavy as a pregnant elephant. It was only capable of straight stitching (without attachments). That said it would stitch through just about anything you could throw at it. Notice the change from Husqvarna to Viking on the next page.

The Husqvarna model 18 was once again based loosely on the Singer 15 principles. However it had many unique ideas, including an easy hold bobbin case. With a simple bolt-on hand assembly or motor, it was very popular with busy sewers. Even today they are still in use. Regularly maintained they will last as long as the metal holds together.

The post WW2 Husqvarna CL 16 & 26 were similar to the 18, a central bobbin Singer 15 design with reverse and a built in light. Again in the beautiful metallic seafoam green and dull mat crinkle green.

The Class 16 or CL16 had a short production run coming out post WW2 and running up until around 1950. It was often in a matt crinkle green. Husqvarna were perfecting their super metallic enamel which they would use for the next few years.

CHAPTER NINE
The 1960s

By the 1960s Husqvarna were on a boom. They had a raft of new models hitting the shops around the world and had patented loads of brilliant ideas.

To go along with their new models they hit new markets. Schools were now a big target for Husqvarna and over the next three decades it paid dividends. There was hardly a school in my area that did not have a few Husqvarna machines in the cupboard ready for lessons.

Most Husqvarna machines were still metallic green. It had almost become their trademark. I can't think of another maker since the Elna Grasshopper that was even close to Husqvarna's super metallic green.

Model numbers were still confusing as they jumped with little reason from model 8 all the way to model 70 and then there were the subclasses, 71A, 19A and so on. Also now there was a superb range of sewing tables available called Delta Tables. These were model 530, 540, 550 & 560. They even sold Husqvarna chairs to go with them.

NOTE: THE NEXT FEW PAGES OF MODELS JUMP UP AND DOWN, SIMILAR TO THE HUSQVARNA MODELS OF THE PERIOD.

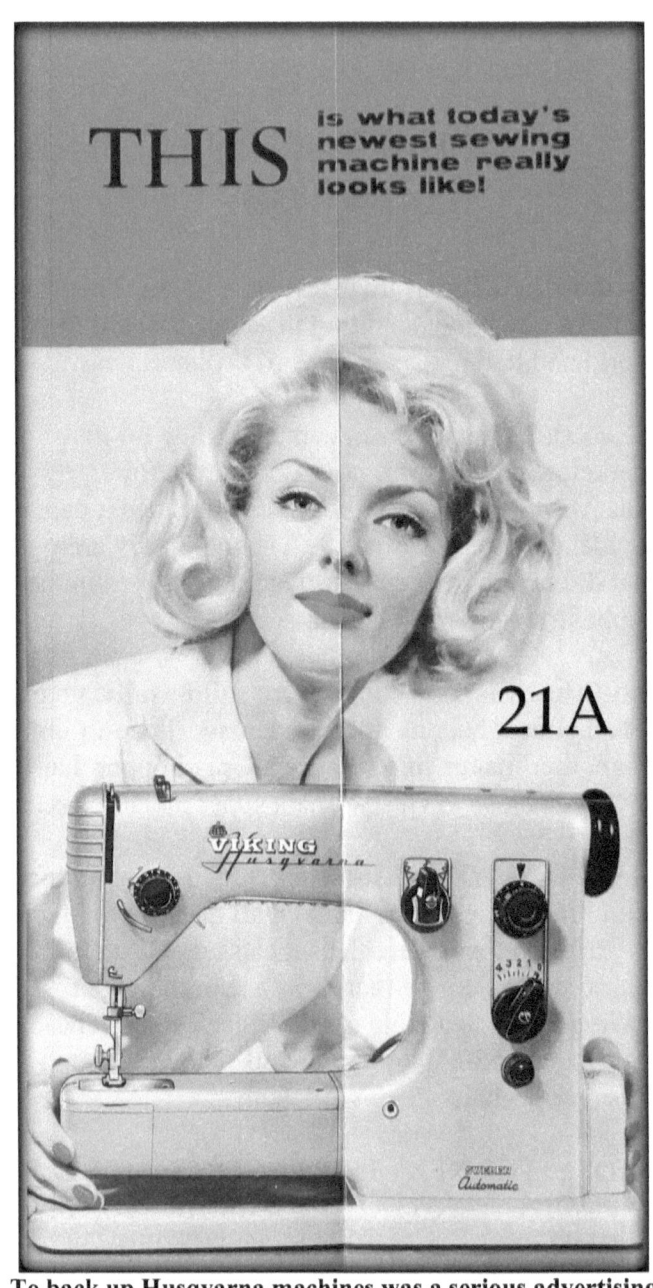

To back up Husqvarna machines was a serious advertising campaign. How 1960s is that look!

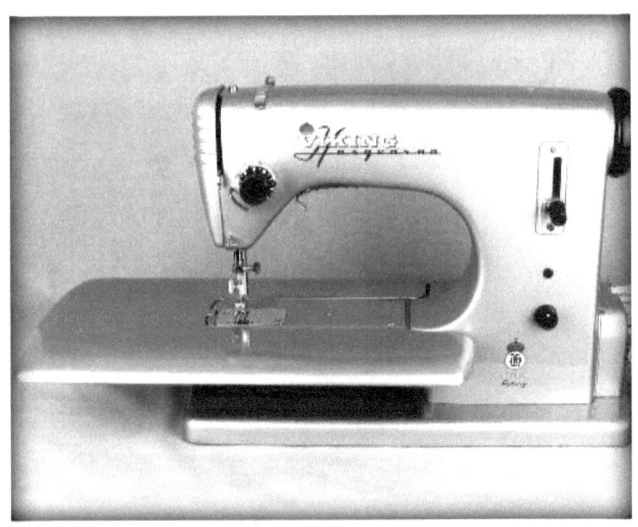

Similar to the Husqvarna model 8 in most ways, the later Husqvarna 71E is just one fab sewing machine. From silk to sacks this baby laughs as she sews and looks great while she is doing it. I always wanted a car in the same shade of light metallic green. It was a basic straight stitch with reverse. It was fast too.

In 1961 the Husqvarna Model 8 was quite dated and still iron. It had the jam-proof hook but it only did forward and reverse which if you think about it machines had been doing since Victorian Times. This was the last 'old style' Husqvarna sewing machine but with a modern rotary hook. It was fast though at 1,500 stitches a minute.

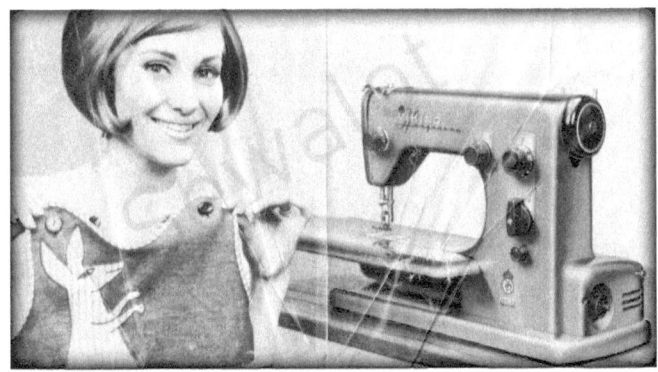

A 60's classic and one of the finest sewing machines ever made. The Husqvarna Viking 19e is almost silent in operation and will sew the finest silk to a leather jacket. The anti-jam hook was sensational and the unique bobbin case was a work of engineering art. Over 150 separate operations went into manufacturing these two. I have a booklet all about tension adjustment on Amazon.

I have never discovered Husqvarna's unusual model system. They jump up and down with their numbers. The iron model 20 (introduced in 1953) had the added zigzag and needle position as well as the new jam-proof hook. It was almost the start of the self-oiling models.

A Little Tip
If you ever have to adjust the tension spring on the bobbin case of a Husqvarna Viking only turn it an 1/8th of an inch at a time and test it. They are very sensitive.

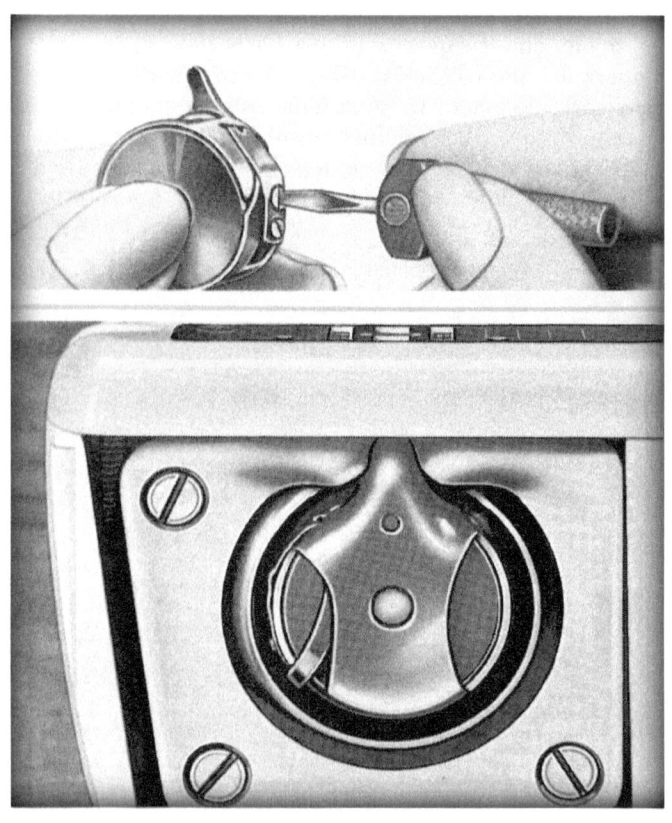

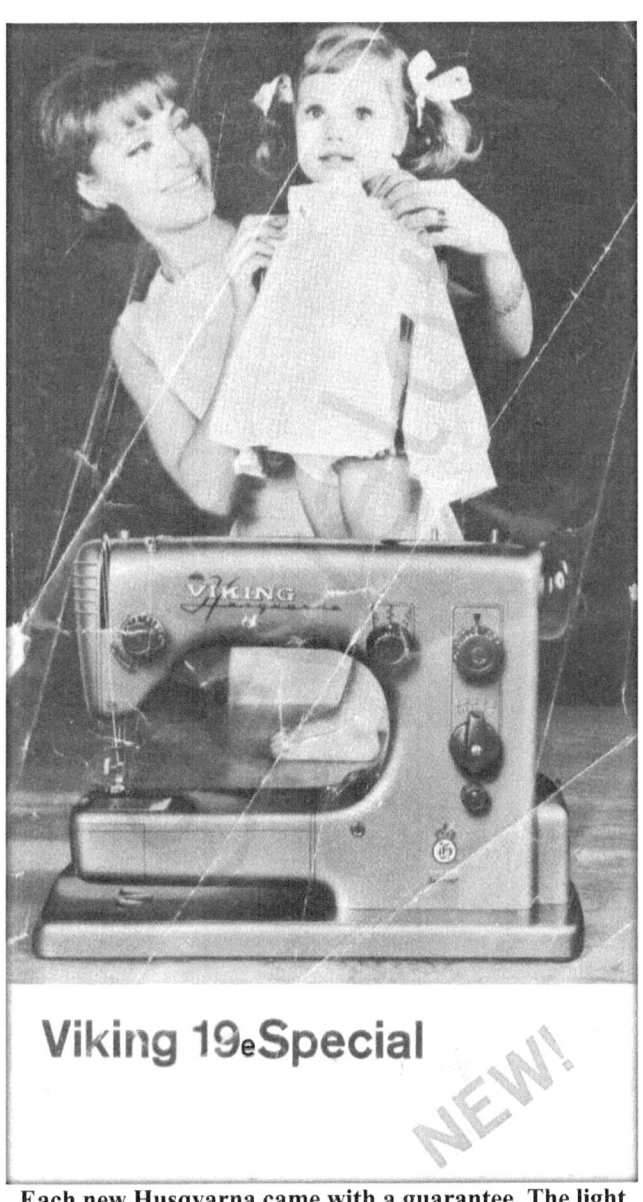

Each new Husqvarna came with a guarantee. The light metallic green models are still widely collected and used today. The 19e ran from 1959 to the middle 1960s.

The Husqvarna model 19E, in my workshop. It is one of the finest sewing machines ever made and came with four patterns. It is highly sought after and collected today. The Husqvarna 19A was thin cast iron to save weight. Dealers would explain the differences in price and performance of each model.

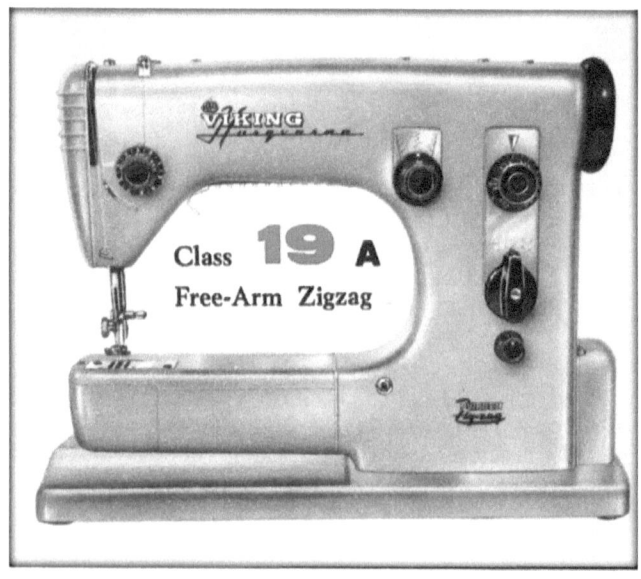

The secret of the 19's success was its price. Made of iron but flexible with a zigzag and free-arm it was selling smack in the middle of the sewing machine price range for 1961. It was slightly slower and lower powered than the CL 21.

The Husqvarna Automatic model 21. Similar to the 19 but with improvements and push-in pattern cams. The model 21 was a stunner, still metallic green but now with a fully enclosed powerful motor and flexible stitches. The CL21 was also available as a basic flatbed and ran from 1955-66.

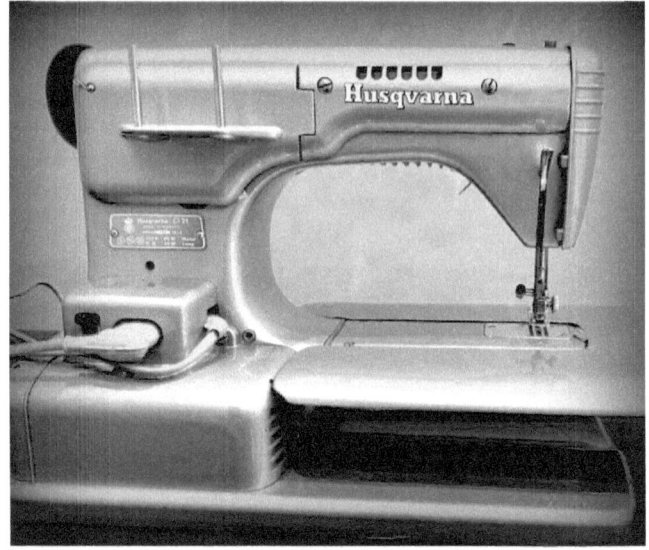

Doesn't the Cl 21 fit that late 1950s era so perfectly! It's sort of space age and futuristic but with classic almost Germanic styling. If this was a car it would be a classic for sure. Open the flap to insert the pattern cams.

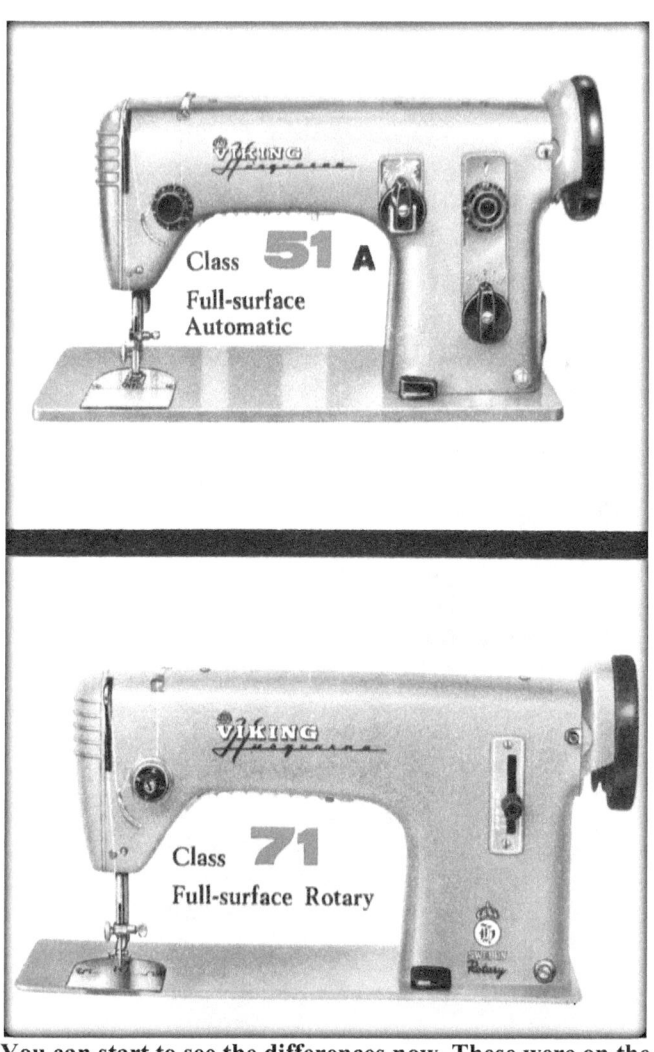

You can start to see the differences now. These were on the same page of a 1961 advertising leaflet. An inbuilt light also lit the work area with almost no shadow. These beauties were the pure Swedish machines. The rise of cheap imports would soon start to put a strain on all high quality manufacturers in Europe.

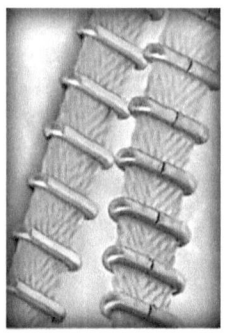

The top and bottom shafts were joined by a toothed fabric belt. It was the precursor of the modern toothed belts of today used widely in industry. On several models there was an easy access swing open panel for mechanical adjustments (just where the cotton pegs are at the back).

The Husqvarna Delta table

The Husqvarna Delta tables looked so 1960s. They had a cutting edge Scandinavian design and are very rare today. Some schools had rows of them. I believe the inventor/designer was Gunnar Nystrom.

A rare Husqvarna Combina 49 flatbed. I always found it tricky getting the bobbin case in easily but set in the Delta-Table you simply reached underneath.

The 1960s were boom years for Husqvarna even though many European companies were on a downward spiral.

One of the reasons was that in America, Husqvarna boosted sales by joining forces with the White Sewing Machine business. The company went on to promote an excellent range of sewing machines across the United States.

Today the Husqvarna name is still on a wide range of domestic goods from fridges and microwaves to chainsaws as well as their world-renowned motorcycles. We'll get to those presently.

Always the trendsetters, for over 50 years, Husqvarna had also been producing space age looking microwaves.

Husqvarna Nordic

The first Husqvarna Nordic sewing machine was based on the popular Singer model 17 design and ran right up until the outbreak of the Second World War with little change.

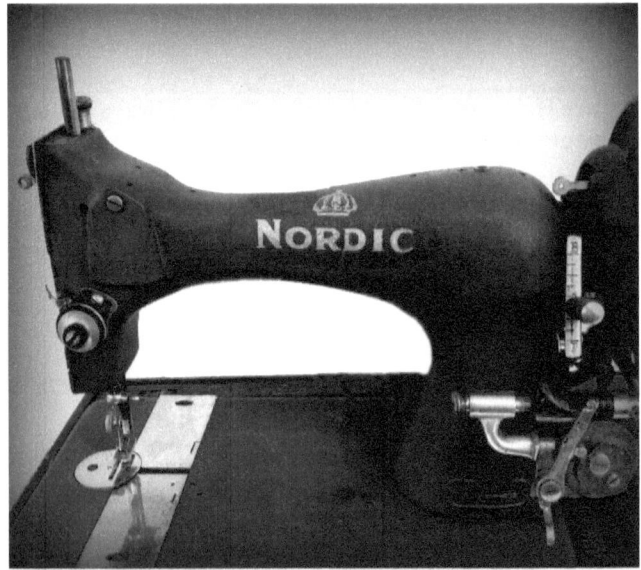

After WW2 improvements kept the Nordic name selling on different models. It was a far longer running machine than many realised which is why I have left it so late in this book. I often wonder if Husqvarna were going to push the Nordic brand name rather than the Viking name that they ended up with. The term 'Nordic' refers to the Scandinavian countries of Northern Europe.

Both these machines were kindly sent in by Keith Hallgren. The Nordic 10 above is an early VS shuttle machine. One of the first 'green machines'. The Nordic Comrie below is a Husqvarna model 15. Comrie were a wealthy family of retailers and distributors in Alberta, Canada, who had originally come from Scotland. They sold Husqvarna, Pfaff and some Japanese models.

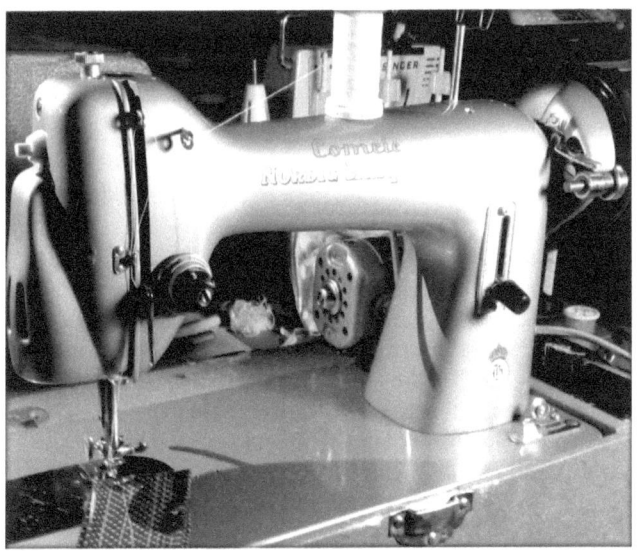

The Nordic name was used on many Husqvarna sewing machines from the earlies vibrating shuttles to the 1961 model 51 Nordic Automatic and finally the Nordic 1390. Many of the Husqvarna Nordic range were flatbed models.

Nordic badged Husqvarna machines changed dramatically from the first vibrating shuttle 19th Century models to the early 1960s with the latest engineering. The Nordic badged Husqvarna machines ran right up until the late 1960s

Here is a later Husqvarna Nordic sewing machine. It is similar to the Husqvarna model 51E (which was a flatbed front loader very similar to the model 21). The flatbed was a bit of a pig to put the bobbin in unless you had nimble fingers. The Nordic name was on several Husqvarna machines for many years. I believe that the very first Husqvarna Nordic sewing machines used the Singer systems and the later the Husqvarna patented rotary hook system (which took industrial bobbins).

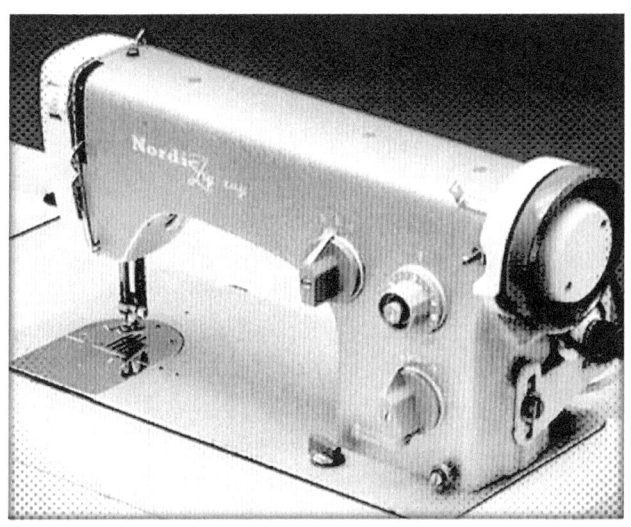

The Nordic Zig-Zag of the late 1950s was so similar to many of the Husqvarna models and sold widely. Almost identical to the Husqvarna 21 flatbed it was set flush into a table. It was made of cast iron and weighed a ton. It was perfect for professional machinists. This is one step below an industrial but more flexible than most. It was capable of almost continuous use and had several oiling points for the main bearings. Motors were easily changed if needed.

Back at the factory Husqvarna were busy improving their machines. At huge expense Husqvarna obtained several worldwide patents and at least 20 Swedish patents for their unique systems. One of the finest was the jam-proof hook mechanism. As I put the finishing touches to this book I believe Husqvarna branded products now have thousands of patents protecting them.

The jam proof hook was a beauty. The hook on a sewing machine grabs the top thread from the needle and in the blink of an eye wraps it around the bobbin case thread to secure a lockstitch. Hooks jam! If you have had a thread jam in a hook you know that one second the machine is stitching along beautifully the next it is seized solid.

In 1953 Allan Herman Eriksson perfected his work on a jam-proof hook. It gave Husqvarna a big leap forward in their marketing and brought many happy smiles to machinists around the world. Allan, a self-taught sewing machine engineer worked for Husqvarna for over 50 years and patented countless ideas. He was an extraordinary sewing machine genius.

A Little Tip

Buy quality thread. I have looked after some of the busiest sewing factories and the finest, most consistent sewing thread I have ever used is Gutermann. None seems to sew better through all fabrics than their 100% polyester. Why ruin all your hard work with cheap thread? For a perfect stitch always use the best.

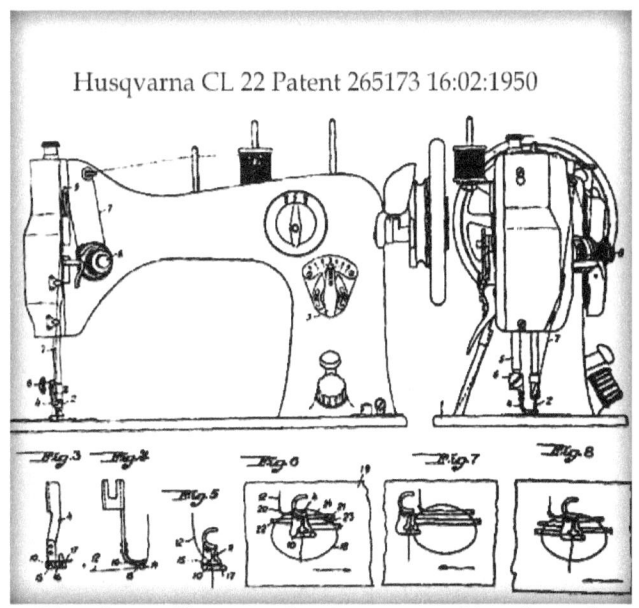

I loved spending countless hours searching through Husqvarna patents. Each one was an insight into a master engineers mind. Genius at work.

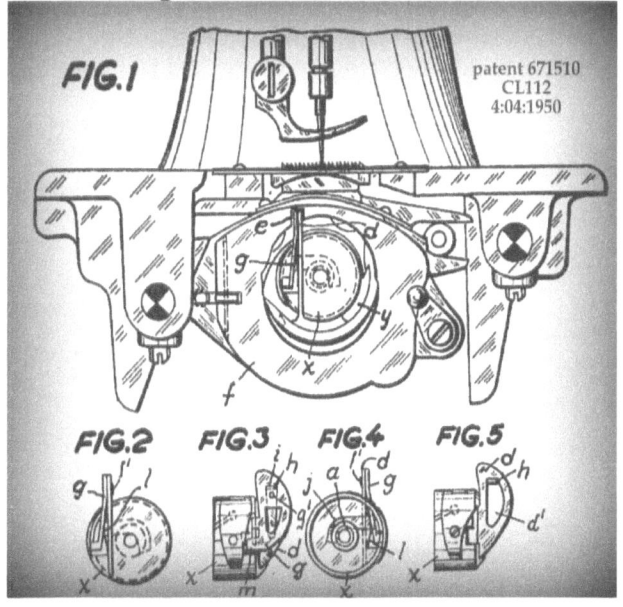

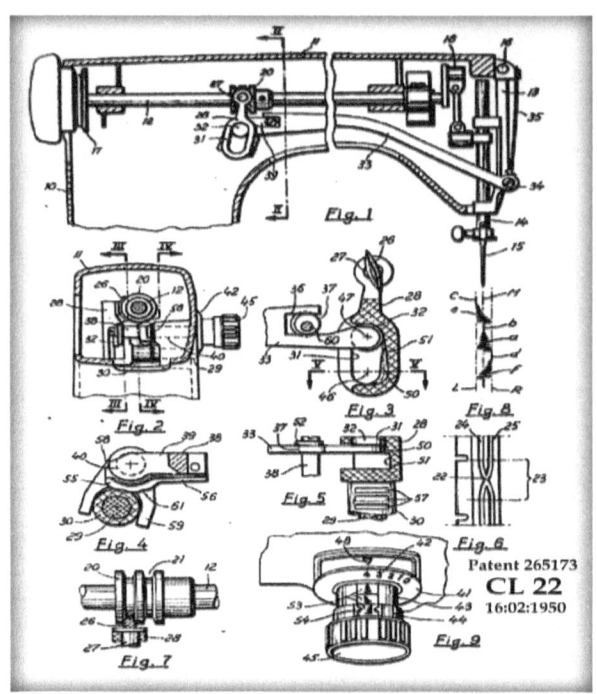

The zigzag cam and gear reduction patents were particularly interesting. Besides Sweden, Husqvarna patented mainly in Germany, the United Kingdom and America.

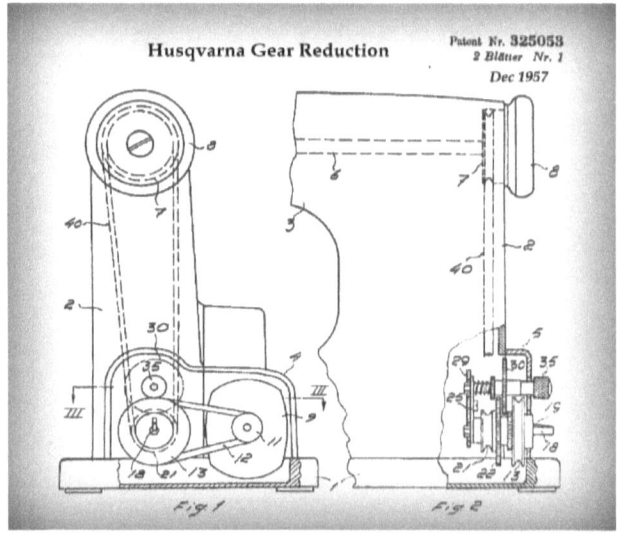

CHAPTER TEN
Viking, South Africa & Canada

I've mentioned earlier why Husqvarna added the Viking name to help sales but there was a little hiccup in South Africa.

I was told that the Viking brand was a trademark belonging to KSIN of Luxembourg. However there is a tale that is worth telling.

In South Africa the Viking name was already in use and trademarked by the Berzack Family. They ran Harrison Sewing Machines from Harrison Street, Johannesburg.

Because Husqvarna could not come to an agreement with the South African business, Harry Berzack informed me that they still cannot officially use the Viking name on machines there. Harry only died recently and it was a great loss to the sewing enthusiast world. I have no idea if the tale he told me was true.

Another weird little anecdote is that in Canada there was a large retailers called Eaton's Department Store. They also had already established the Viking brand name on their machines (apparently bought from the Riccar in Japan). Because sales for Husqvarna Viking machines were low in Canada at the time, Eaton's used the name until the stores closed. They also used the Viking brand name on many of their household appliances.

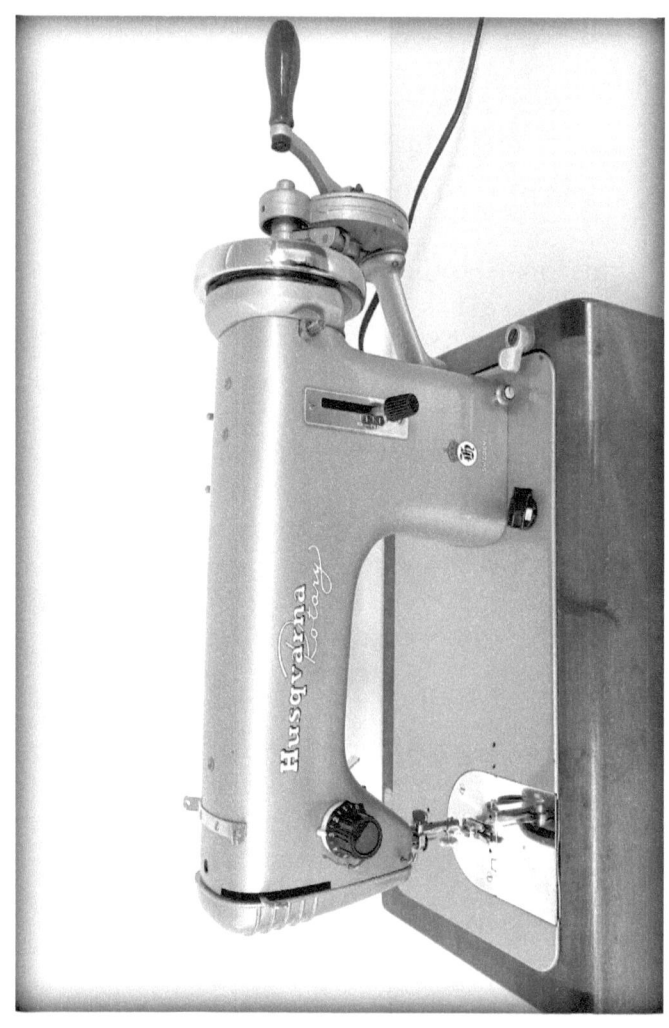

As far as I am aware the last 'green machine' by Husqvarna was the hand driven model 71E in 1966 and the first of the 2000 models. This 71E was kindly sent in by Hans C. Endrerud. It was mainly sold as electric and had a full rotary bobbin. Notice the oil holes on top. These would soon disappear on all the later models as Husqvarna's patented self-oiling bearings came into production.

Now let's get back to our amazing story. One of the turning points in sewing machine history was the achieving of self-lubricating bearings by Husqvarna. By the early 1970s they had perfected bearings that could be sealed for the life of the sewing machine without ever needing oil.

The 1970s was a new era for Husqvarna. They had survived the onslaught of cheap imports by diversifying and keeping their quality high. They were on a boom in North America and elsewhere.

New models were flourishing as were the colours. Out was the old Husky green that had stood the firm in good stead since WW2. In was a kaleidoscope of 1970s modern shades.

Colours from deep burgundy to shocking orange were displayed in shop windows across the world. All these bright and beautiful machines are now highly collectible.

The company logo also modernised. If I get enough money I'll do the Kindle version of this book and that will be in colour, so you may get an idea of what I am talking about.

The logo still boasted the royal crown, now enveloped in green laurels as a nod to their old green machines. The first Husqvarna Viking model 2000 machines were the last green machines before turning white.

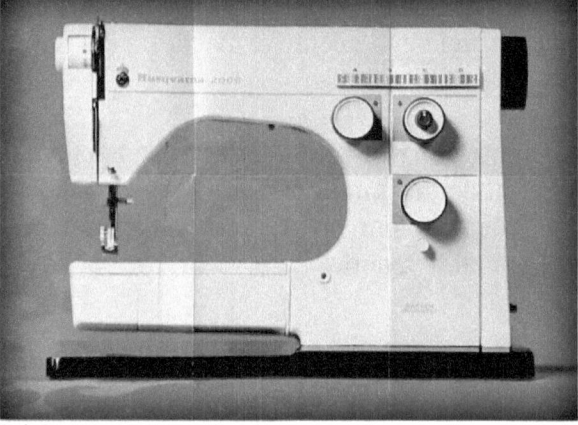

Bright new colours for the latest range. I used to love collecting the old sewing machine advertising leaflets when I was a kid. I had no idea half a century later I would be digging through countless filing cabinets pulling them back out. The almost maintenance free Husqvarna 2000 is a joy to sew with. It proved hugely successful with the schools and colleges here in Britain.

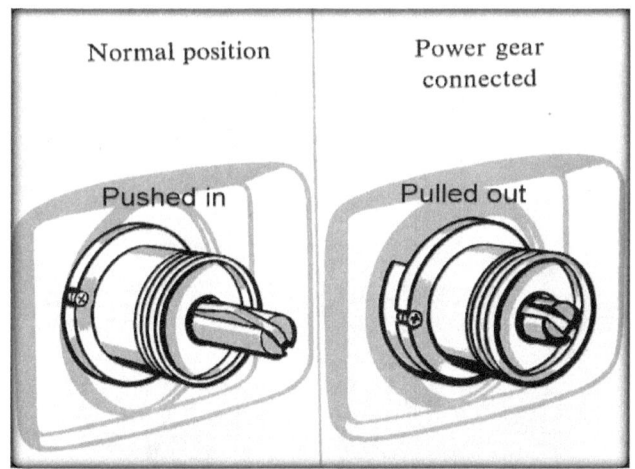

Its powerful motor had a brilliant reducing gear which, when engaged, dropped the machine speed but increased the penetration power through heavy fabrics.

A Little Tip

If you are teaching someone how to sew on a Husqvarna engage the speed reducer. It allows the machine to go as slow as a sleepy snail.

I'll tell you something funny. When I was a kid I thought it would be a good idea to buy a couple of spare Husqvarna motors. A lifetime later I still have those motors, brand new 'old stock', never used. I have never needed to fit them. Husqvarna motors are amongst the most reliable in the world.

As an engineer, self-oiling steel is as close to a magic trick as it gets. Let's have a closer look.

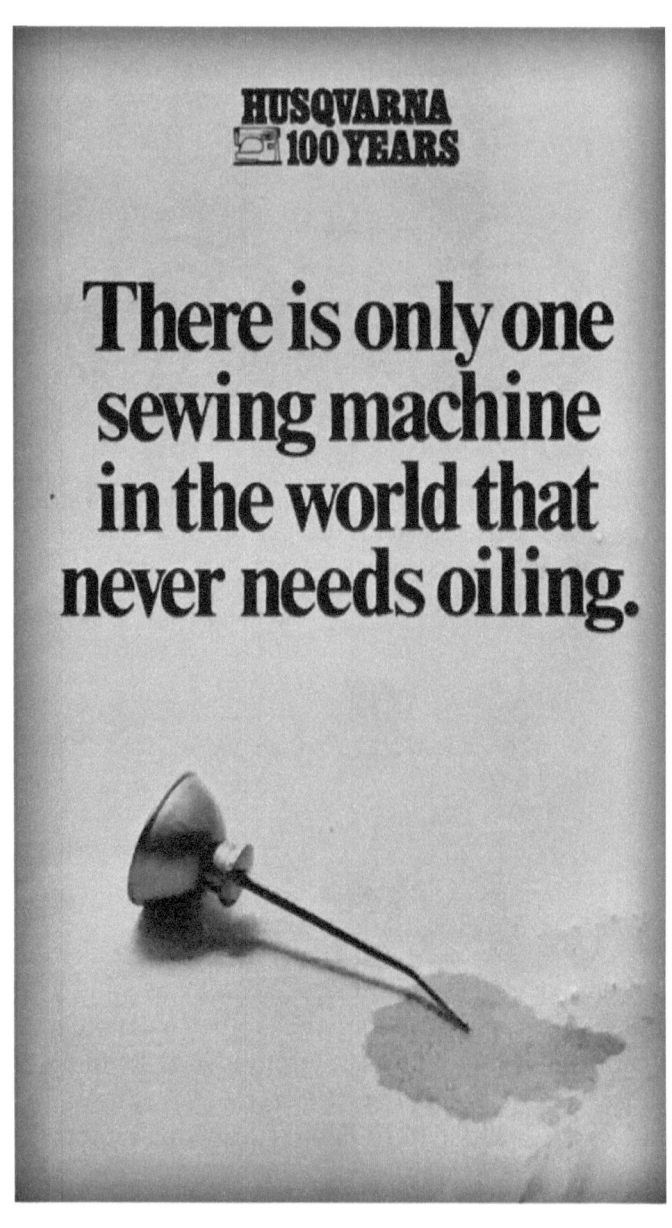

Husqvarna were one of the pioneers in self-oiling, low maintenance machines. Although it was on earlier models, in 1972 they proudly started advertising their latest range of oil free machines worldwide.

Although sintered steel had been around for many years it was Husqvarna that perfected and patented its use as bearing material in sewing machines. Other companies used cast iron that had similar properties with the ability to absorb oil. This made their machines very popular with schools in Europe for their durability and toughness.

If you have an old Viking Husqvarna they love being used. If they sit idle the bearings, and many of the internal parts that have been designed to move, simply lock up. Locked bearings on any machine are a nightmare. Use it or lose it!

Take your old friend out of the cupboard (the machine that is) and give it a good run. You will see it come back to life. There is nothing a Husqvarna enjoys more than being used.

The first models that I am aware of Husqvarna advertising their self-lubricating bearings on was the 2000 series around 1972 (100 years after their first sewing machine). However sintered steel and oil impregnated bearings were already in regular use.

The Husqvarna bearings were soaked in oil. When the machine revolved the friction warmed the bearing and it released some of the oil onto the sewing machine shafts. When the machine stopped and cooled it absorbed the oil back. Self-lubricating Husqvarna bearings were either sintered steel or phosphor bronze around the take-up lever assembly. If your take up lever bearing does fail it is repairable as parts are still made to fit.

A Little Tip

Due to the use of this special (and expensive bearing material) not using your Husqvarna will allow the bearings to lock. Regular use, bringing the machine up to working temperature (30 minutes of regular sewing) will extend the life of your machine dramatically.

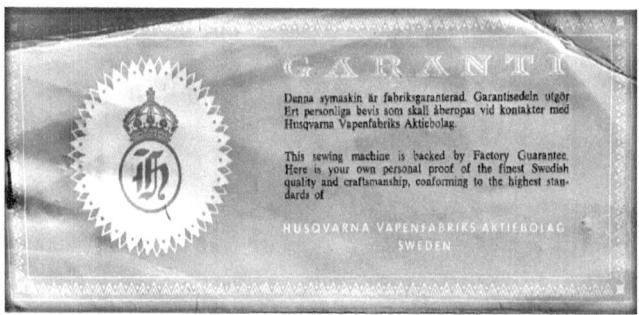

Precision design, expert craftsmen, Swedish steel and constant production standards, with examinations at all stages of manufacture, produced machines second to none.

Every machine came with a sample of the stitching that it made, and the number of the person that set it up at the factory. The Husqvarna guarantee was backed up by dealerships around the world.

The machines up until now were the last mainly-all-metal Swedish Husky machines that were built to last a lifetime. Unhindered by deteriorating plastics, many of the machines can still sew like new.

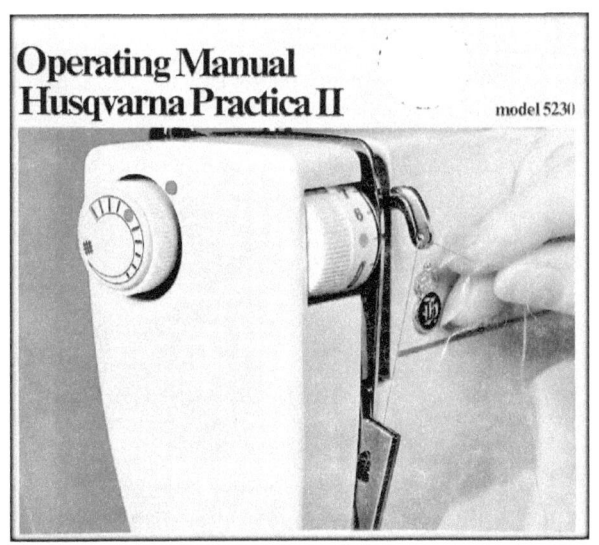

The Husqvarna 5000 series and the Practica series were almost identical. Around most of the world the Viking Husqvarna was sold as the Viking 5000 series but in the United Kingdom the similar machine was sold as the Husqvarna Practica, and Practica II. The Practica name had been used on earlier models from 1968 onward.

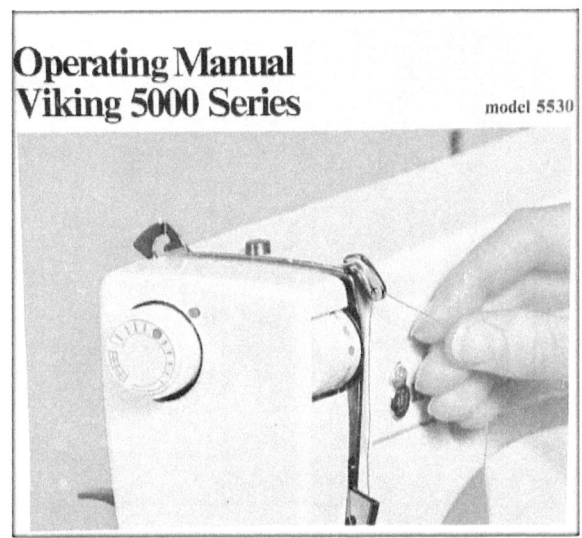

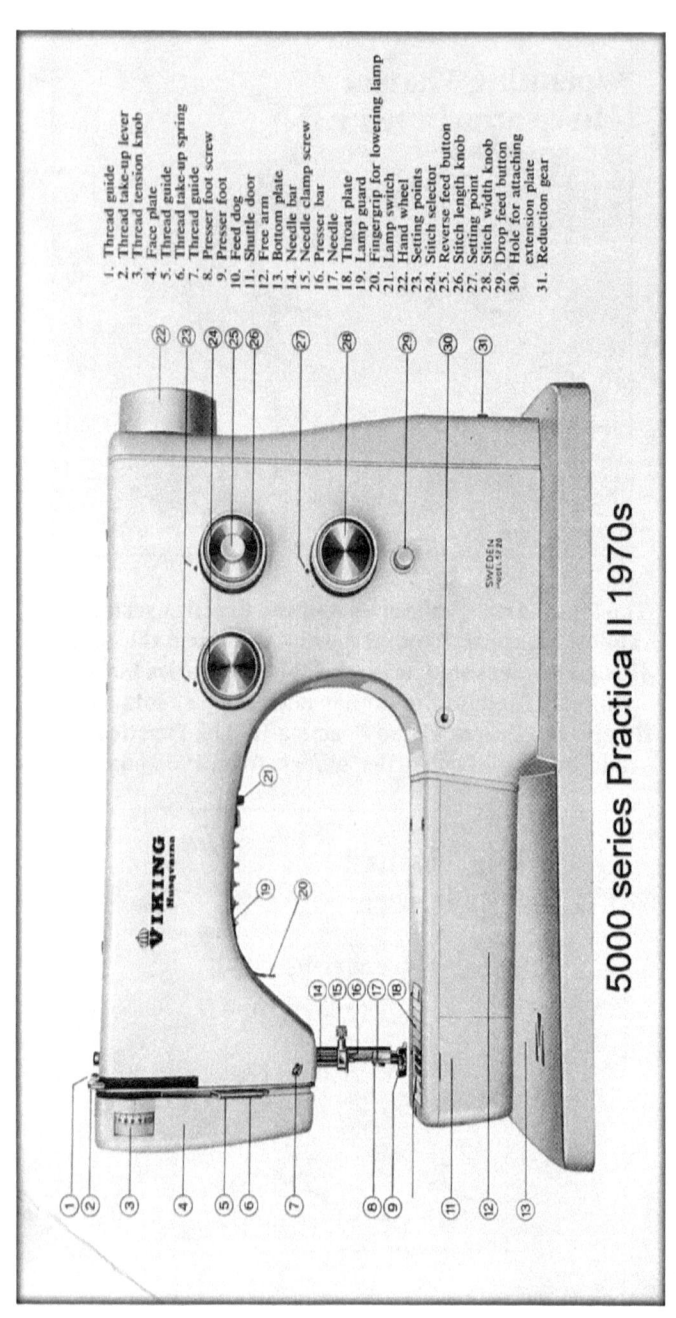

5000 series Practica II 1970s

1. Thread guide
2. Thread take-up lever
3. Thread tension knob
4. Face plate
5. Thread guide
6. Thread take-up spring
7. Thread guide
8. Presser foot screw
9. Presser foot
10. Feed dog
11. Shuttle door
12. Free arm
13. Bottom plate
14. Needle bar
15. Needle clamp screw
16. Presser bar
17. Needle
18. Throat plate
19. Lamp guard
20. Fingergrip for lowering lamp
21. Lamp switch
22. Hand wheel
23. Setting points
24. Stitch selector
25. Reverse feed button
26. Stitch length knob
27. Setting point
28. Stitch width knob
29. Drop feed button
30. Hole for attaching extension plate
31. Reduction gear

These similar models caused quite a headache for our local Eastbourne dealers in South Street called Ryders, and poor old Varney (the main school suppliers in the area) in Portslade. He nearly blew a gasket every time a school argued that they wanted the 5200 Husqvarna machines not the Practica II.

Incidentally these top quality machines came at a price. Now back in 1977, I was 20 years old and working at the family firm. I was getting £60 per week minus all the usual deductions. I came home with about £50 a week. Most of that seemed to disappear in bills. Born free taxed to death. Was that on a gravestone?

The Husqvarna Viking model that was shining to me out of Ryders window was £180. That related to over three weeks wages back then and two years spare cash. A small fortune to me. I looked up the average wage in the UK in 2022 and it was around £600 a week (nice for some). That relates to the same machine costing around £1,800 today.

Now you can see why they were so good. All that expensive technology and quality came at a price. However, like my dad always told me, nothing lasts like quality. The fact that so many early Husqvarna Viking sewing machines are still being used around the world proves that case perfectly m'lord. I don't expect you could build many rubbish mountains from old broken Husqvarna machines.

Husqvarna were now promoting the Viking name on its sewing machines and the name Husqvarna was reducing in size. However the 6000 series still had the royal crown, now it was above the V of Viking rather than the H of Husqvarna. The 6000 series was not vastly different to the earlier 2000 models and sold worldwide. The 6000 was available in royal red and plain white.

A few years later Husqvarna, who had been trading as Husqvarna AB (limited to us) updated their Swedish crown to a modern logo.

A little tip
Many people don't know that the Husqvarna logo apparently still depicts the end of a gunsight as viewed from the barrel. How clever is that.

In the 1970s Husqvarna were producing some of the finest sewing machines around. The 3000, 5000 & 6000 range were available in this decade. As well as the Husqvarna Vanessa and others. They all ranged from about £140 -£200 depending on the amount of stitches. All these machines were professional models capable of constant work. Most of them did a huge 6mm straight stitch which was great for bigger work and basting. Their hardened steel teeth seemed to be able to pull just about anything through the machine.

Of course the machines went wrong. I mean humans were involved after all! To maintain the efficiency of Husqvarna Viking machines, dealerships around the world were trained and had complex workshop manuals. I've had to repair more than one Husqvarna that had tried to fly of the sewing table, especially at schools. The last one I repaired was pulled of the table by a helpful husband going to open the curtains. He tripped over the power cable, launching the machine into space. He carefully put everything back and took the dog for a walk! Strangely enough he happened to be out when I arrived to repair the machine a week later. I explained to his wife what I thought must have happened. A quick phone call to the squirming guilty party confirmed it. I never did find out what transpired when he got home!

Several of the models had extra patterns available in what Husqvarna called their 'Colormatic System'. These were pop-in patterns cams pushed into the back of the machine. Each cam added an extra four patterns to the machines arsenal. You could buy endless designs from Husqvarna.

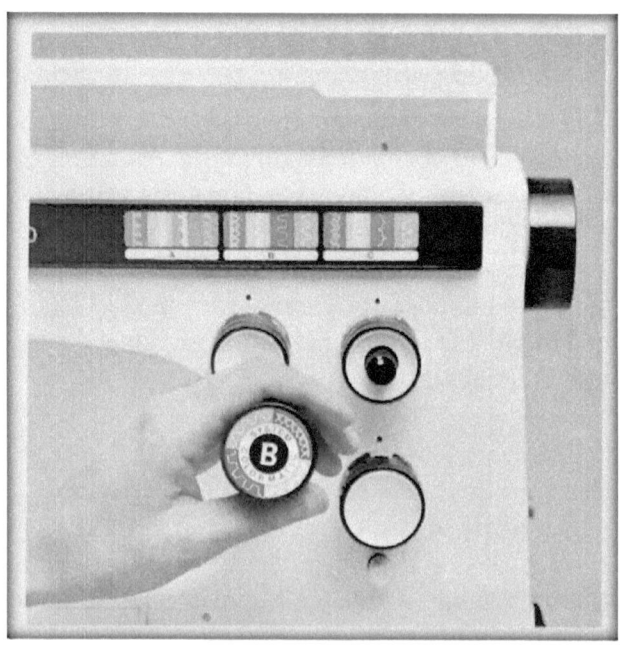

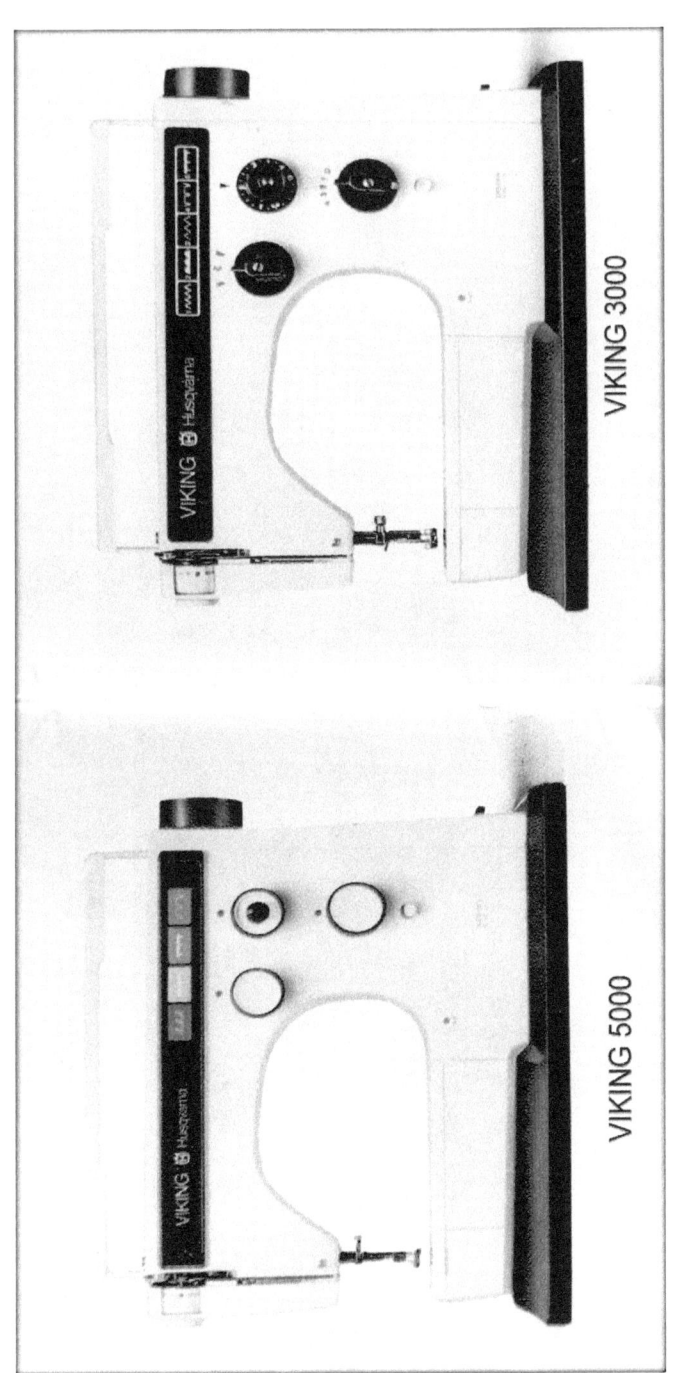

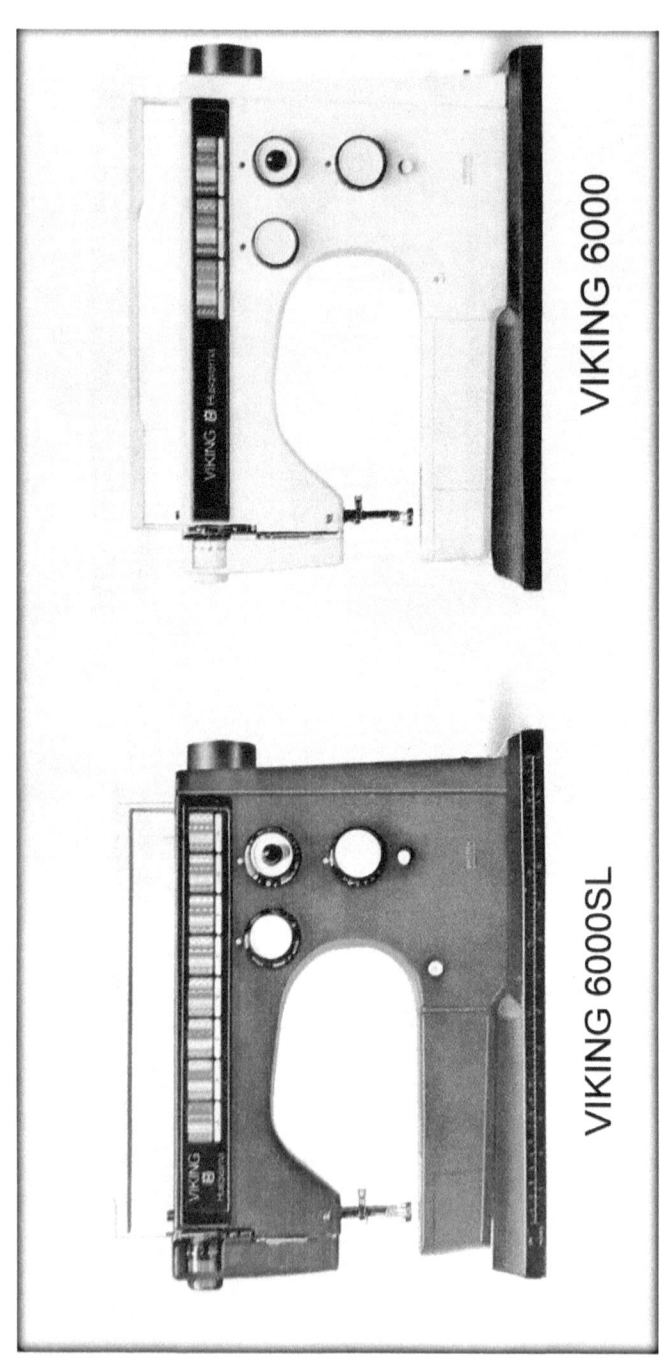

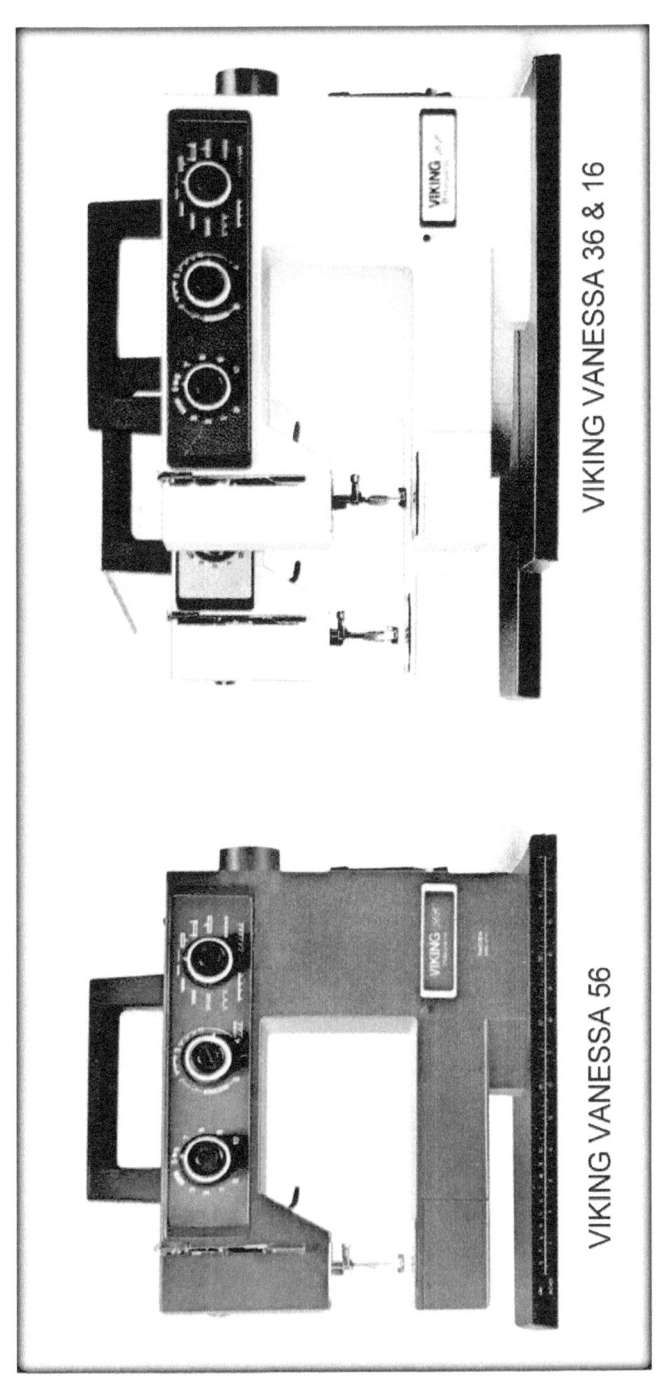

CHAPTER ELEVEN
All Change

In the 1970s a host of new models flourished at Husqvarna. No one seems to know why the model numbers went ballistic. For a 100 years they had been slowly and methodically going up with their model numbers. It was reasonably easy to date the machines not by serial numbers but by models, but then it all went stratospheric.

In 1977 Electrolux, the huge Swedish white goods supplier, purchased Husqvarna. They held Husqvarna as a subsidiary of the Electrolux Group for a period.

Electrolux sold off many of the company's manufacturing arms, including the motorcycles. Husqvarna had been producing motorcycles since 1903.

The motorcycle division was sold to the Italian makers Cagiva and in 1987 it became part of MV Agusta Motor S.P.A.

The only way to travel in 1903

I believe that they also sold the powerful Husqvarna name for other merchandising. It was now known all over the world. Possibly in 1997 the Husqvarna sewing machine brand name was marketed separately. We'll get back to that presently.

The 1980s saw the fabulous Husqvarna Optima range. These were a series of machines to suit all needs. Although Husqvarna had pioneered firstly electric, then electronic, then fully blown computer sewing machines, this was more of a sideways step. Easy to use machines with a limited number of manually selected patterns.

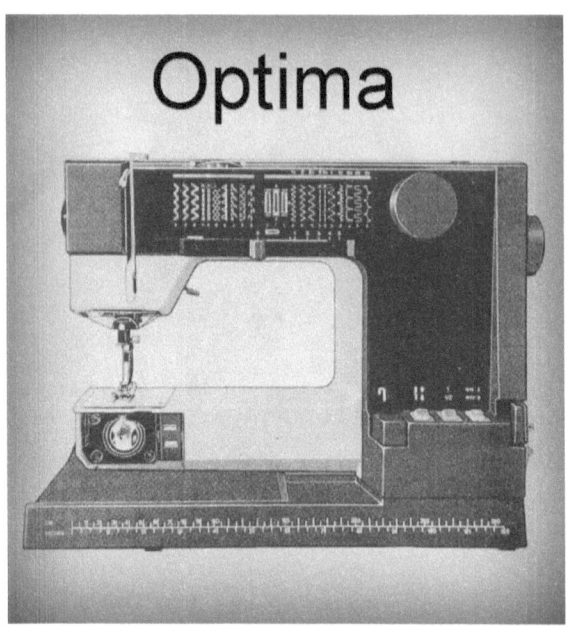

Husqvarna were now naming their machines again and pushing the Husqvarna badge as much as they had the Viking badge. The Optima range were incredibly advanced inside but looked simple outside. Pretty much the perfect machine. I loved them because a single screw held the back panel. Access for repairs was fast. I have never come across an Optima that wasn't repairable.

A point of interest with the Swedish name cut-up and sell-off is, I believe that in BMW had a stake in the Husqvarna motorcycle side from 2007 and their bikes have been manufactured in several European countries. In 2013 Husqvarna motorcycles (in some form) came full circle back into Swedish hands, possibly as part of KTM motorcycles.

As a motorcycle rider for most of my life I seriously thought about putting more information in the book. However I knew it would add at least another 100 pages and drive a few of my enthusiasts wild. So I will just include possibly the most famous Husqvarna motorcycle ever, the Husqvarna 400.

The Husqvarna 400 came out in 1969. It was a two-stroke off-road powerhouse built on a lightweight frame and went like a rocket. One was bought and ridden by the famous American Actor Steve McQueen.

1971 Sports Illustrated front cover.

In 1971, the August Issue of Sports Illustrated had one of the most iconic images of the star ever taken. Steve McQueen used his bike in the closing sequences of the film, On Any Sunday. The bike instantly became legendary.

It boosted the sales of Husqvarna motorcycles and sent them skyrocketing. When the bike was sold in Las Vegas on the 10th January 2013 at the Bally's Hotel & Casino by Bonhams Auctioneers, it fetched a mouth-watering $230,500. WOW.

Interestingly Husqvarna bicycles, which started life in 1896 and stopped in 1962, are now also back with a great range of bikes including electric ones! I love my electric bike. I converted my Holdsworth back in the 1980s and have modified and ridden it ever since. Mind you I would love to ride a new Husqvarna electric!

Now, back to business. In 1986 Electrolux acquired WCI (White Consolidated Industries). Like I mentioned, White's had been selling sewing machines in America for decades, including Husqvarna models. However they had been hit hard by cheaper imports and were ripe for a buy-out by Electrolux.

The constant movement of the Husqvarna brand is all very puzzling isn't it! We will deal with the final piece in our jigsaw puzzle later. Unions, mergers, subsidiaries and buy-outs are not what I wanted to write about so let's get onto something very special indeed.

Our old Swedish weapons maker was about to make a final bang in the field sporting world.

CHAPTER TWELVE
The Last Sporting Gun 1989

In 1989, just a short 300 years after the original Husqvarna Vapenfabriks (Husqvarna Weapon Makers) was formed, the last sporting guns were made. It was the same year that the super Husqvarna model 610 was launched proudly boasting a sticker with 300 years on it.

What an amazing achievement in world history, one company that has survived for three centuries. Many of their machines carried this badge in 1989.

Right up until the 1960s Husqvarna were still making sporting guns, pistols and rifles. In Sweden to this day they are highly regarded and collected.

By the 1960s Husqvarna were the largest selling high-power lightweight bolt-action rifles in the world. Superb Swedish steel, European walnut stocks, and flawless workmanship from centuries of progress had all combined to make perfect weapons. Each rifle was marked with the gunsmith's signature and no finer example existed. Countless championships were won with rifles like these.

I believe that earlier, back in 1969, the small arms were relocated from the Husqvarna Group to Forsvarets Fabriksverk at Eskilstuna. The two companies had been working together ever since collaborating on the Swedish Armies assault rifles.

Back at the Husqvarna factory, just a handful of highly skilled staff were kept on to repair and maintain the weapons that came back for renovation. When time allowed they would hand-build a few sporting guns.

In 1989 all this came to an end with a special commission of just 15 shotguns. They would be hand built by Bertil Granqvist and Nils Abrahamsson. The stocks were made by Petrus Ruckman and the astonishing engraving was carried out by Hans Svensson (who had been engraving Husqvarna guns since 1949).

These super special shotguns became known as 'Jubilee' models. Although each one cost over £13,000 at the time, all were sold before manufacturing was even commenced. I would love to know how much one of them was worth today.

Interestingly a weapon maker's bench looks very similar to a jeweller's bench which looks pretty much identical to my work bench. Just rows of specialist tools all within easy hand-reach of the central work area.

CHAPTER THIRTEEN

A Little Tip
Husqvarna Viking machines love Schmetz needles. They would put a pack in every sewing machine they sold. For years they used Schmetz exclusively. If you want the best stitch on your Viking Husqvarna stick to Schmetz, they are simply the best. No I'm not an agent for the company but I do buy them and fit them in every machine I service.

New Husqvarna badged machines in our modern world are vast and varied, coming from many sources, so I will leave my story shortly with what I believe to be one of the last great Swedish Husqvarna sewing machines.

Not the fabulous orange monster they made in the 1970s (which is now retro cool by the way) but possibly the best modern Swedish sewing machine ever made, the Husqvarna 500 computer. I don't rate it as the best because it was Husqvarna's most complex or advanced machine but for simple perfection in sewing.

Husqvarna had great success with their amazing Husqvarna 1, a fully blown computer machine with simply remarkable abilities. The Husqvarna 1 used their patented Omni Stitch. What's that? I can hear you ask. Well it is mind blowing. Since the invention of the four motion feed by A B Wilson way back in the mid-19th Century, little had changed in the way work moved through a sewing machine.

The four motions of the feed were simple, up forward, down and back. This remains basically the standard movement for most machines. However the Omni Stitch not only did the four standard movements but in total the feed dogs moved eight different ways. This allowed fabric to move whichever way you set the Husqvarna 1, allowing wonderful embroidery stitches and patterns to be created.

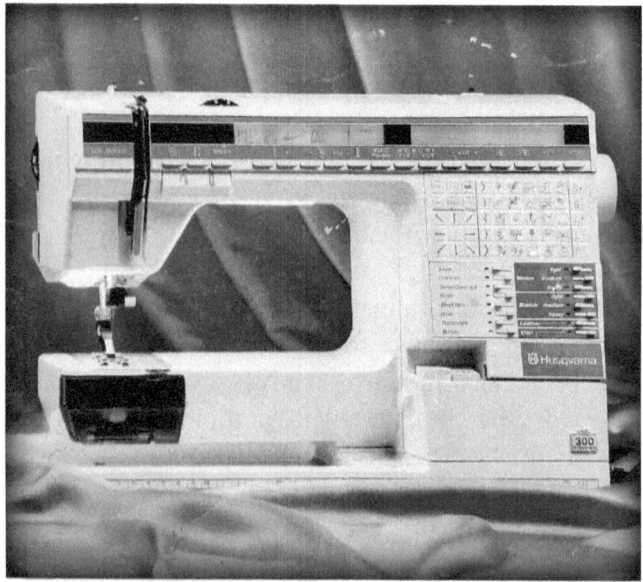

This 1992 Husqvarna 1 (1200) moved the work not only forwards and backwards but left and right too! The computer could create all sorts of wonders.

Now, what they did with the Husqvarna 500 was strip away all unnecessary 'fancy' stuff leaving you with an easy to use semi-professional machine. It was the successor to the bestselling Husqvarna 400, a powerful metal machine encased in a friendly paint free casing.

I remember Bengt Gerborg, (who had become President of Husqvarna in 1985) explaining that the new 500 combined all the best ideas they had to offer in a competitive machine (and boy it sure did). Touch button control, drop in anti-jam bobbin, automatic stitch adjustment (so that when you pressed a certain pattern it also automatically set the perfect width and length for that particular pattern).

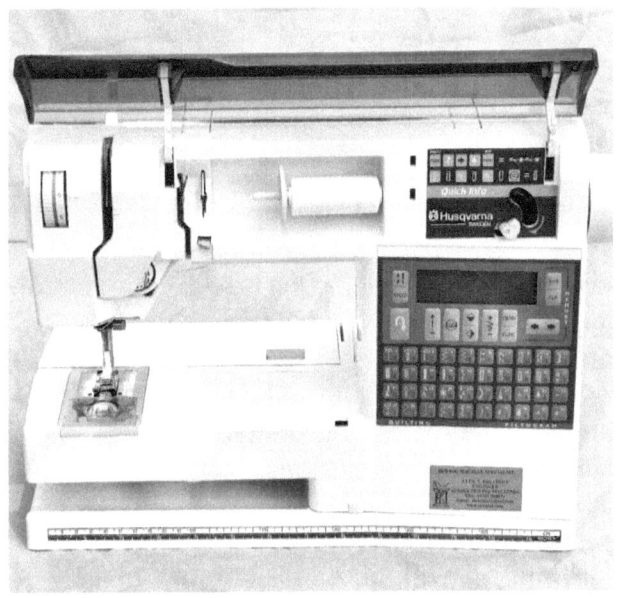

The Husqvarna 500 Computer was introduced in 1994. See what I mean about their confusing model numbers, 20 years earlier we were on model 2000 so we seem to be going backwards again.

It all sounded amazing and it was another year before I got to sew on one. I wasn't disappointed. It was a joy to sew with.

Over the next 25 years I would come across many Viking 500 machines. They appealed to home sewers and professional machinists alike. They all loved sewing with it.

There had not been a machine so widely adored by all its users since the stunning little Singer Featherweight and the Singer 201 (which my last book was on). Its secret was that it made extraordinary complex stitching so easy anyone could use it. To me it was near the pinnacle of perfection in sewing.

In 2002 President Svante Runnquist spoke to fellow members of ISMACS (International Sewing Machine Collectors Society). They were on a tour of the factory to see the museum and the latest computer manufactured machines.

Svante told the group that the new machines (still hand finished) were working to higher tolerances than a fine watch. My old friend Maggie Snell, the ISMACS treasurer and the all-encompassing glue that bonds the society together, told me how she watched the finished machines being packed for export.

At the time around 550 workers were still involved in the manufacture of sewing machines at Husqvarna and they were producing around 675 sewing machines a day in Sweden.

Today the Husqvarna name is still on a superb range of sewing machines from basic sewers to incredible embroidery machines.

They also have an impressive Museum open to the public at Hakarpsvägen, Huskvarna, Sweden. It is well worth a visit and set in one of the most beautiful settings you will ever see.

CHAPTER FOURTEEN

You might want to skip this last chapter. I was trying to keep to sewing machines and guns but let me try and explain the name today. There's no guarantee it's correct, the details seem to change depending on who you Google or talk to.

Because the Husqvarna brand name had become synonymous with Scandinavian quality it had a large 'brand' value. You could put Husqvarna on just about anything and it would sell. By 2006 the Husqvarna name was on countless products.

I believe that in 2006 Electrolux moved on the Husqvarna brand. But I could be wrong here as it all gets amazingly confusing (unless you have a wonderful legal brain). Anyway, Husqvarna becomes independent again and is listed on NASDAQ OMX, Stockholm Stock Exchange.

Husqvarna (Viking Sewing Machines VSM AB, Huskvarna, Sweden) was bought by Kohlberg & Co and later merged or changed hands to become part of SVP, Singer, Viking, Pffaf. Singer was bought in 1999, Husqvarna and Pfaff in 2006. The Husqvarna Group still remains as Husqvarna AB.

Now, as I write we can see separate businesses, brand names and umbrella companies all circled around the Husqvarna brand name, from motorcycles to bicycles, lawnmowers to chainsaws.

The Husqvarna Group are the world's largest producer of outdoor power products and world leaders in several other industries. The brand name just keeps on growing and is available in countless countries across the world.

On the sewing machine side, in January of 2010 SVP (Singer, Viking and Pfaff) announced that they were stopping Swedish sewing machine production. Some design and development was kept in house and coordinated with other R&D units around the world. New models now come out of the latest super-modern Far East factories. It was the end of nearly 140 years of continuous sewing machine production in Sweden.

Although I've been told that the original Swedish Husqvarna factory does not currently produce sewing machines, who knows what the future will hold.

SVP Worldwide is an American private company and are the largest sewing machine suppliers on the planet. They supply a range of machines for all tastes from fabulous and fancy to plain and simple.

I believe that Platinum Equity may have acquired SVP Worldwide. Wow, isn't big business confusing!

The Husqvarna brand name has always been synonymous with quality. Today it can be found on everything from work boots to massive ride on lawnmowers that cost as much as I earn in a year. Interestingly the very first powered lawn mowers also cost a year's wage in 1910.

Well, here we are at last. Originally I had planned to do just the early Husqvarna machines but it has been a fun time travelling through the centuries.

What an amazing journey, from a small group of villagers (that used to burn their own huts and flee each time marauding Danes approached), to computer sewing machines. We have spanned the centuries and been on a whirlwind tour of one of the oldest and finest companies on planet earth. It was complicated but I hope you thought it was worth the effort.

Hopefully one day Husqvarna will investigate and bring to life more of their amazing engineers and give them the publicity they deserve. Who knows they may even drag out some sewing machine serial numbers!

Well that's it folks, another book in my Sewing Machine Pioneer Series all done and dusted. Bye for now.

Although Singer, Viking and Pfaff are now sold under the SVP banner, I believe that in 2015 the Singer Red 'S' cameo badge and 'Singer' name were still exclusive trademarks of the Singer Company Limited, Sàrl.

Husqvarna
The Early Years

By
Alex Askaroff

Sewing Machine Pioneer Series

On The Road Series

There are seven books in Alex Askaroff's **On The Road Series**. They cover his working life around Sussex encompassing a world of stories from the ages.

Book One: Patches of Heaven

Book Two: Skylark Country

Book Three: High Streets & Hedgerows

Book Four: Tales From The Coast

Book Five: Have I Got A Story For You

Book Six: Glory Days

Book Seven: Off The Beaten Track

"If you read any of Alex's 'On The Road Series' you will read them all. They are totally addictive, beautifully crafted and wonderfully inspiring."
Eliza Cooper

Alex Askaroff at Birling Gap

For collectors and enthusiasts
of antique sewing machines and great stories
why not visit

www.sewalot.com

For other publications By Alex Askaroff Visit Amazon

Isaac Singer
The First capitalist
No1 New release

Most of us know the name Singer but few are aware of his amazing life story, his rags to riches journey from a little runaway to one of the richest men of his age. The story of Isaac Merritt Singer will blow your mind, his wives and lovers his castles and palaces, all built on the back of one of the greatest inventions of the 19th Century. For the first time the most complete story of a forgotten giant is brought to you by Alex Askaroff.

No1 New Release. No1 Bestseller, Amazon certified.

*If this isn't the perfect book it's close to it!
I'm on my third run through already.
Love it, love it, love it.
F. Watson USA*

Patches of Heaven is the first book in Alex's popular 'On The Road Series'. We start Alex's working life and follow him as he earns his living. With Nine No1 New Releases on Amazon, Patches of Heaven with enthral readers of all ages.

Alex's second book in his hilarious 'On The Road Series' continues with his travels. We meet forgotten war heroes and crazy customers by the bucket load, from the 1930s debutant balls at Buckingham Palace to a sailor who had a lucky escape from HMS Hood, before its encounter with the formidable Bismarck Battleship.

Once again Alex has worked his magic. These true stories will have you in stitches, you may even shed a tear but you will be left with a happy heart.

Tales from the Coast, Book Four in Alex's On The Road Series, continues the true stories which he brings both England's history and people vividly to life. The stories are as pleasurable as a warm bath after a long day. From the disappearance of Lord Lucan in Uckfield to the Buxted Witch, from William Duke of Normandy to Queen Elizabeth's Eastbourne dressmaker, Tales from the Coast is crammed with a fascinating mix of true stories that will have you entranced from start to finish.

Yet again Alex has woven his magic. I kept saying I never knew that and I'm a local. This may just be one of the best books I've ever read!
J. Vincent

Alex, I've read every book James Herriot ever wrote, and my favorite topics in his books are about the animals and the meals, just like my favorite stories in your books are the ones that talk about your experiences working in people's homes. I love them. Thank you so much.
Joe Edmiston
Louisville, KY

Alex Askaroff has had Nine No1 New Releases on Amazon. For decades Alex has been enthralling readers around the world with his writing. Off The Beaten Track is the seventh and final book in his 'On The Road Series' and completes his working life before retirement.

The 7th Duke of Devonshire's dream started in the middle of the Victorian Era and continues to flourish to this day. World renowned author Alex Askaroff tells the story of Eastbourne in his own unique style, reviving long forgotten characters from the town. We meet local families, fishermen, smugglers, kings and queens, ghosts and even an old witch. It is a tale not to miss.

www.sewalot.com
Alex's No1 antique sewing machine site.

Sir Sewalot, protector of Sewalot.com

Husqvarna Timeline

1. During the 1300s there was a well maintained fortress called Rumlaborg near the area of Jönköping and the mill houses known as Kvarnhus, Husquernen. The area by the falls became collectively known as Husqvarna near Jönköping.
2. The fortified encampment attracted craftsmen and weapon-makers. A thriving community grew.
3. News of the craftsmen reached Denmark. Vikings attacked the settlements looking for weapons and raw material. Any craftsmen would also be taken if caught.
4. By the 1500s the area around Husqvarna and Jönköping was a thriving fortified military community, including craftsmen and their families.
5. By the 1600s the unconnected communities had come together to harness the power of the rivers, not only for grinding grain but weapons manufacture. Farming had turned into industry.
6. 1620, King Gustav II orders the area to be organised under one central command.
7. As a weapons centre, the Swedish Crown could now place orders with the newly formed, Jönköping Rifle Factory.
8. 1688, improved mill works allowed for a massive expansion in manufacturing, under the control of Count Erik Dahlbergh.
9. 1689, the new rifle factory receives its first big order from the Swedish Crown. Husqvarna is officially born.

10.   Output in one year rose from 1,100 to 12,000 rifle barrels.
11.   1757, the Swedish Crown relinquishes its hold on Husqvarna allowing it to become a private company. The Husqvarna gevärsfaktori (Husqvarna Rifle factory) is born.
12.   Around 1810 the spelling of the town changes from Husqvarna to Huskvarna. Husqvarna, the company, is now a unique and separate from the town. It is this divergence that eventually leads to one of the largest brand names in the world. A tiny acorn had been planted.
13.   1845, almost all weapon manufacturing activity in the area is now centred around Huskvarna.
14.   1850, over 1,000 people are employed at Husqvarna. In the same year the arsenal was privatized under the ownership of Fredrik Ehrenpreus.
15.   1855, a massive injection of capital sees the latest improvements in manufacturing techniques at Husqvarna. The Industrial Revolution is underway.
16.   1867, Husqvarna becomes a limited company. On the 22nd of November the first AGM is held.
17.   1869 sees Husqvarna receive one of its largest weapons orders, but a period of peace in Europe is on the horizon.
18.   1870 sees a sudden drop in orders and cancellations. The area falls into recession. Husqvarna drops to 170 employees.
19.   Hugo Tamm, who focuses in taking ailing companies and resurrecting them, steps in. Hugo

already owns part of Husqvarna and looks to diversification from weapons for the first time. The acorn has sprouted shoots.

20. 1872, the first ever sewing machine, the North or Northern Star is produced at Husqvarna. Although a failure in performance it was a huge success on the sales side. It shows Husqvarna that there is a big market for sewing machines.

21. 1874, kitchen equipment is made for the first time.

22. 1877, Wilhelm Tham is promoted to president of Husqvarna by the entrepreneur and industrialist Hugo Tamm.

23. 1883, The Freja was the first sewing machine that made a near perfect stitch for the Husqvarna Company. The name ran until 1925.

24. Husqvarna make a series of sewing machines and carry on until the present day. They become world leaders in innovation and design.

25. 1889, Husqvarna start transferring from steam and water power to electricity. This transforms production techniques. Hydroelectric power is later developed along the rivers.

26. 1896, Husqvarna manufacture their first bicycles.

27. 1903, Husqvarna manufacture their first motorcycles and later go on to dominate world sports. The acorn has become an oak.

28. 1903 is the same year that the CB, Round or Oscillating Bobbin machine is produced by

Husqvarna. This is the start of their 'modern' machines.
29.   1919, the first lawn mowers are made. Husqvarna go on to develop the most advanced lawn cutting equipment in the world. They grow to become the largest supplier of grass cutting equipment on the planet.
30.   1934, The CL 12 CB-N model is introduced, production exceeds 41,375 machines in 12 months.
31.   1939, Husqvarna invests in a small 'peoples car' but it fails to materialise.
32.   In 1946 Husqvarna has over 6,000 employees. The entire area has prospered due to the factory. The company are producing nearly 73,000 machines a year.
33.   In 1947 Husqvarna produce their first zigzag machine called simply Zig-Zag.
34.   1950, from this period Huskvarna the town merges with Jönköping.
35.   1951, Husqvarna pass the 100,000 machines per year mark. They are now serious competition to makes likes of Singer and Pfaff.
36.   In 1953 Husqvarna produces the superb CL 20 series with its jam-proof hook and low maintenance.
37.   1959, Husqvarna produce their first chainsaws. They go on to produce the finest chainsaws in the world.
38.   1960s, Husqvarna rifles win countless awards and trophies and become the largest selling high-powered lightweight rifles in the world.

39.   1960-61 also sees the introduction of the revolutionary 2000 series of oil-free multi-stitch sewing machines.
40.   1966 The Husqvarna Viking CL21, first made in 1955, is still selling but is discontinued.
41.   1968, Husqvarna produce their first portable angle grinders and cutters.
42.   1969, the small arms division is relocated to Forsvarets Fabriksverk at Eskilstuna.
43.   1970, Lil Wettergren, Sweden's first ever board member, is elected to the board of Husqvarna. It is a 'glass ceiling' moment for women's liberation in Sweden.
44.   1971, Husqvarna patents the sealed-for-life, self-lubricating bearings and introduces the 6030.
45.   Husqvarna becomes Husqvarna AB or limited. The Swedish Crown logo is also modernised. The latest logos is the end of gun barrel outline.
46.   1972 sees the Husqvarna Centennial Series launched exactly 100 years after the North Star.
47.   1978, Electrolux acquires Husqvarna.
48.   1979 sees the first use of Husqvarna computerised step motors to produce unlimited stitch formations by moving the needle bar and feed in micro-movements.
49.   1980 the Husqvarna 6690 has the ability to stitch the alphabet. Janome and Singer are also producing computer machines in a race for technology. A fully blown embroidery machine will soon follow.

50.   1989, Husqvarna produces their last sporting shotguns with a special edition of 15 handmade beauties. All are sold before they are finished and become priceless.
51.   1990, the amazing Husqvarna 1 series is launched with the new 'all direction' computer controlled feed. It is a WOW point in sewing machine history, a precursor to small embroidery machines.
52.   2006, Husqvarna becomes independent again and is listed on NASDAQ OMX in Stockholm.
53.   2010, Swedish production of Husqvarna machines stop at Huskvarna.
54.   2013, A Husqvarna 400 owned by Steve McQueen sets a new world record for an off-road motorcycle at $230,500.
55.   2015, SVP, Singer, Viking, Pfaff. The Singer Red 'S' cameo and 'Singer' are exclusive trademarks of the Singer Company Limited Sàrl (Société à Responsibilité Limitée).
56.   2021, The Husqvarna Group continues to expand exponentially worldwide. Multiple companies, under the Husqvarna brand become the largest suppliers of outdoor products in the world.
57.   2022, the Husqvarna brand name has outgrown the country of Sweden. Husqvarna has become a marketing phenomenon. The tiny acorn has become a giant oak with a thousand branches across the world. Its amazing story continues…

Writing my book on Husqvarna started in 1989 and finished in 2022. It is my hope that I have added many of the missing pieces of this incredible brand so that others may carry on the torch.

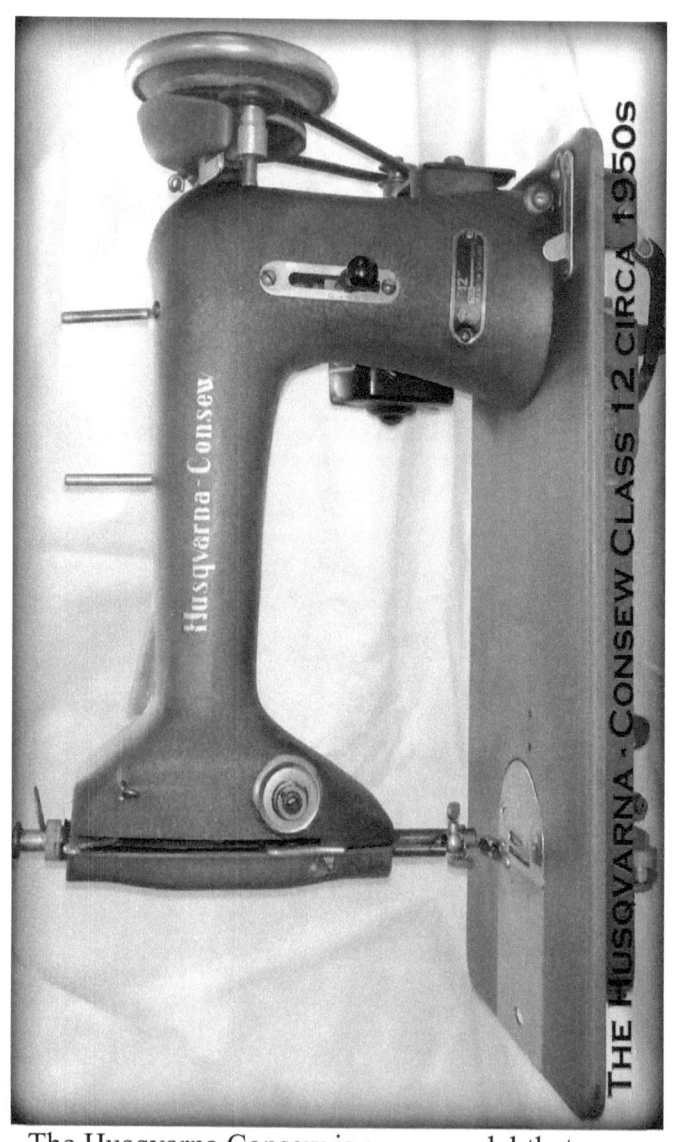

The Husqvarna Consew is a rare model that came out in the 1950s. It is actually a Husky Class 12 flatbed. Many thanks to Cathy Boyer.

www.ingramcontent.com/pod-product-compliance
Lightning Source LLC
Chambersburg PA
CBHW020426220526
45464CB00002B/584